Japanese Goldfish

An imported ryukin. From a photograph by Dr. R. W. Slinfeldt of a specimen at the Bureau of Fisheries, Washington, D.C. The parts which appear black are vermilion red in life.

Japanese Goldfish: Their Varieties and Cultivation

A practical guide to the Japanese methods of goldfish culture for amateurs and professionals

HUGH M. SMITH

Water-color drawings (from life) by J. Urata

WAKING LION PRESS

ISBN 978-1-4341-0384-0

Published by Waking Lion Press, an imprint of The Editorium

Waking Lion Press™, the Waking Lion Press logo, and The Editorium™ are trademarks of The Editorium, LLC

The Editorium, LLC
West Jordan, UT 84081-6132
wakinglionpress.com
wakinglion@editorium.com

Contents

v

Introduction

There exists in the United States great and growing interest in the keeping and cultivation of Japanese and other races of goldfish. This interest is one phase of the noteworthy amount of attention now devoted by young and old to the study and care of living creatures, and is aided by the facility of obtaining at reasonable prices desirable material for stocking aquaria and ponds.

The pleasures and profits of raising goldfish are destined to be experienced by many more people; each year thousands of men, women, and children begin to exhibit an interest in this subject by seeking to learn where and how to secure the goldfish. and how to raise and care for them. The demand for the fish keeps far in advance of the available supply, and there is thus created a need for more extensive cultivation and more establishments where goldfish are produced.

This little work is presented in the belief that, while American goldfish culturists and fanciers have developed most efficient methods, there is much for professionals and amateurs to learn from the Japanese, and that a discussion of goldfish and their culture from the Japanese viewpoint will be of practical value and general interest to Americans.

While there have been published many works dealing with the goldfish and their cultivation. there is none that has just the scope and viewpoint of the present one; and it is believed that there are herein set forth certain aspects of Japanese methods that have never before been made known to western readers.

The data on which this work is based were obtained by the writer during two visits to Japan when, under the guidance of the leading Japanese fishery and fish-cultural authorities, the principal goldfish breeding establishments were inspected and information was Obtained at first hand from their proprietors. Supplementary information has been derived from published papers by Professors Matsubara, Mitsukuri, and Watase. The accompanying colored illustrations of ten varieties of Japanese goldfish are made from paintings from life by Mr. J. Urata in Tokyo, and are herewith reproduced through the courtesy of Prof. S. Matsubara, Director of the Imperial Fisheries Institute, Tokyo, by whom they have been copyrighted; these plates accompany his interesting paper published by the Bureau of Fisheries.

WASHINGTON

October 1, 1909

1 General Aspects of Goldfish Culture in Japan

Importance of the Goldfish to the Japanese People

The Japanese are the leading breeders of goldfish; their methods are the most original and successful; their varieties are the most beautiful and interesting.

The goldfish occupies a prominent place in the daily lives of the Japanese. Among the creatures kept for purposes of ornament and amusement—monkeys, birds, musical crickets, singing frogs, etc.—none are in such general demand or are employed in such large numbers as the goldfish; and probably in no other country are any non-useful animals maintained by a larger proportion of the population than are goldfish in Japan.

Interest in goldfish is manifested by all ages and in every class of society, from the humblest peasant to the highest court official. The small boy on a holiday will be made supremely happy by the purchase of a goldfish costing half a cent while a wealthy connoisseur may give one hundred and fifty dollars for a single pair of fish of select breed.

Immense numbers of goldfish are sold on fete days, and children are the largest purchasers. Goldfish vendors, carrying their ware in wooden tubs suspended from a shoulder bar, mingle with the crowds in the parks and on the streets or station themselves at points of vantage and display their living toys to the passing throngs.

It is stated that in feudal times in Japan, even in years when famine prevailed and hundreds were dying of starvation, the demand for and the trade in goldfish continued with but little abatement, because the children craved the fish and their demands could not be resisted.

The vogue that the goldfish acquired in Japan many years ago and has retained with increasing popularity is an index of a significant feature of the Japanese character. The love of the purely beautiful pervades all classes of people, and is evidenced in many ways that are either unknown or but little developed in many other nations. It is very natural that the esthetic temperament of the Japanese should find much to gratify it in the beautiful colors and the graceful forms and movements of the goldfish; and it is noteworthy that of the two oriental peoples among which the cultivation of the goldfish reached an advanced stage at an early date, the Chinese should have directed their efforts mainly to the production of the grotesque, bizarre, or horrifying, while the Japanese strove for the graceful, harmonious, and pleasing.

In the Japanese homes, goldfish are usually kept in small globes suspended in rooms or in balconies, or in ponds or fountains in the miniature landscape gardens with which a large proportion of the houses are provided.

In the thousands of landscape gardens, parks, and temple grounds all over Japan, there are ponds and lakes stocked with turtles, carp, and goldfish; and one of the favorite amusements of the crowds that constantly resort to such places is to feed the turtles and fish. Just as in Venice there is always a vendor ready to supply one with corn for the doves of Saint Mark, so at the public resorts in Japan there is always a person to provide hollow balls of colored rice flour to be thrown into the ponds. The balls are light and for a few minutes float like corks while the fish push them about with their noses in the efforts to eat them; after a time, becoming water soaked, they gradually disintegrate, sink, and are devoured.

The goldfish is a common theme in Japanese decorative and industrial art, and is a favorite subject for biological investigation. Some of the leading men of science of Japan have delved into the natural history of this fish, and have written most entertainingly of its various phases. Being a plastic material, the goldfish when skillfully bred, yields many surprises to the biologist as well as to the cultivator.

Origin of the Goldfish and Its Culture in Japan

Many things that have been firmly established in Japan for centuries in reality had their origin in China, and among the more noteworthy of these is the highly colored cultivated variety of goldfish. The goldfish is Possibly native to Japan, and fish having the dull coloration and simple form of the original wild species are found in open waters all over Japan, but in some cases these are as likely to have been the progeny of fish that escaped from private ponds and reverted to the wild type as to have been natives. At any rate, there is no evidence of the existence of the brilliant, cultivated fish prior to their importation from China.

The history of the introduction is lost in obscurity, but it appears to be established that as early as the year 1500 some goldfish, probably of the simplest variety, were brought from China to a town near Osaka; and many other importations were doubtless made in early times from China and Korea, where the cultivation of this fish must have begun at a very remote period.

The cultivation of goldfish in Japan began several centuries ago, and had attained considerable extent long before the founding of the United States as a nation. It seems that as early as the first decade of the eighteenth century, a breeder of goldfish began business at Koriyama; and the author has visited at that place a goldfish farm that was started about 1763 and has been in continuous operation to the present time. This establishment was at first maintained only for pleasure. but later became a commercial enterprise and has for many years been conducted at great profit.

The introduced variety of goldfish like various other things that the Japanese obtained from outside their country, was vastly improved upon as a result of independent methods of culture applied at a very early date; and new varieties were soon developed that are still being cultivated.

Centers of the Industry

Goldfish are bred for pleasure or profit all over the Japanese Empire, and it is only in the most northern island, Hakkaido, where the cold is intense, that successful culture is impossible.

The chief centers of the industry are the great capital city of Tokyo which, with its two million people, offers a superior market for all kinds of goldfish in addition to having a temperate climate most conducive to successful culture; Koriyama, a small place near the ancient capital cities of Nara and Kyoto. which also has excellent marketing facilities and a salubrious climate; and Osaka, the Venice of Japan and the second city of the Empire.

Koriyama is the most important center, and has about three hundred and fifty goldfish breeders whose annual output is upward of ten million fish. At some establishments as many as six hundred thousand are produced and sold annually, while at others the yield may be only a few thousands.

Tokyo and Koriyama may be said to be the headquarters of two different schools of goldfish culture, with different breeding methods, different standards of excellence, and different fashions in fish. The Tokyo school dominates the northeastern part of the Empire, and the Koriyama school holds sway over the western part of the main island of Hondo and the islands of Shikoku and Kyushu.

2 Japanese Goldfish Breeds

The Wild Fish

The wild fish from which the multicolored and multiformed varieties of goldfish have been produced is a very plain species, with nothing to suggest the wonderful possibilities of development which it has undergone. The moderately elongated and compressed body is covered with large, coarse scales; the head is unsealed and smooth; the fins are relatively small, and the color is uniform olivaceous. The normal length is eight to twelve inches.

The goldfish was originally placed in the same genus as the domesticated Asiatic carp, and was named Cyprinus *auratus* by Linnaeus. It differs, however, from the common carp in having no barbels, and in having the pharyngeal teeth in a single row on each side; it has therefore been put in the same genus as the crucian carp or karass, of European waters, and its proper scientific name is *Carassius auratus*, which literally means the golden

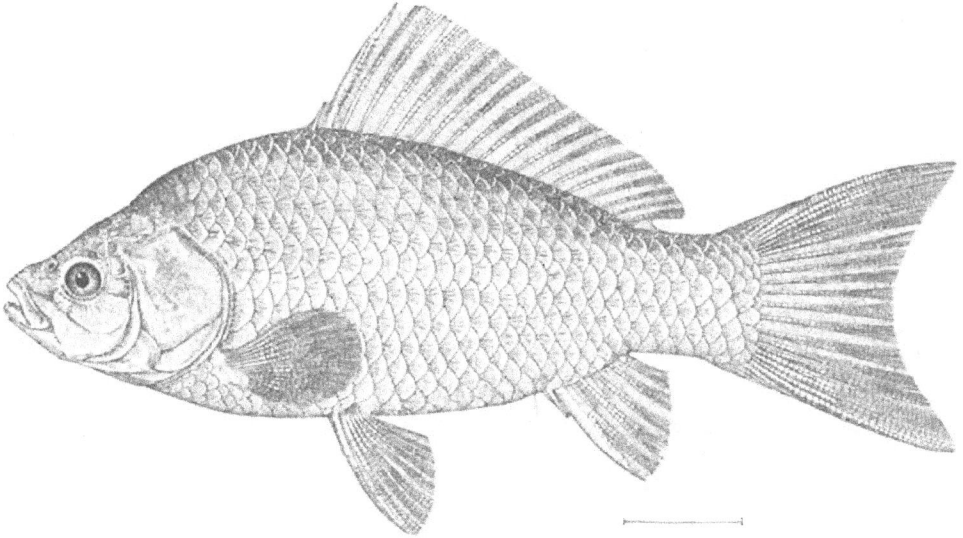

The wild goldfish

or gilded karass. The goldfish is sometimes not inappropriately called the gold carp, but this name is not distinctive because a golden variety of the common carp is now extensively cultivated.

The original home of the fish was China. Authorities do not appear to be in accord as to whether the species was native to Japan, where it is now widely distributed, and it may be that this point may never be conclusively determined.

Evolution of the Varieties

In a group of fishes of the carp family related to the goldfish there is a tendency to albinism; and doubtless the parent stock from which all the cultivated varieties have sprung was albino or partial albino. A deficiency of dark pigment in the skin of the wild goldfish would leave a whitish, yellow, or golden color;

while irregular distribution and concentration of the dark pigment would result in a variegated coloration, with blackish or
dark greenish spots or blotches separated by yellow, golden, or
whitish areas. By the selection of such abnormally colored fish
for breeding purposes, light and variegated races were in time
established. Abnormalities in form may have arisen in and been
similarly reproduced from wild fish, but most probably these
arose in the course of the cultivation of already established or
incipient color varieties. From these primitive departures from
the normal, all of the extraordinary variations in form and color
that we now possess have been produced, after hundreds of
generations, by selected breeding.

One of the ablest American biologists and embryologist, the
late Prof. John A. Ryder, called attention to the fact that the
varieties of goldfish "are the most profoundly modified of any
known race of domesticated animal organisms." In the course
of a paper published in 1893 he discussed the origin and significance of some of the modifications, and advanced the interesting
theory that the greatly enlarged fins in some of the varieties is
correlated with a degeneration of the muscular system through
disuse, owing to "their continued restraint in small aquaria for
many generations." The feeble and almost totally deficient swimming powers of certain varieties are said to have been "purposely
cultivated by oriental fish fanciers." and the energy that would
have been expended "in the production of motion of the body
in the water has reacted in other ways upon their organization,
and especially upon the growth of the fins." In the elaboration
of this theory, Ryder suggested that the enlarged fins may serve
as supplemental respiratory organs, the caudal in particular
being very richly supplied with capillaries and often presenting
an enormous surface for the possible exchange of gases: and
he asked whether this hypertrophy of the fins may not have

"been developed partially in physiological response to artificial conditions of respiration . . . in the restricted and badly aerated tanks and aquaria in which they have been bred for centuries."

It is not necessary to discuss the foregoing views, but it should be remarked that the statements regarding the breeding of Japanese goldfish in badly aerated or restricted aquaria and tanks are entirely erroneous, and any theory based on such an unwarranted assumption is untenable, for, as will hereafter be seen, the Japanese have never raised goldfish under such conditions, and the salient features of the various kinds of ponds in which they have for generations been hatching, rearing, and holding their fish are the ample space afforded and the most perfect oxygenation of the water.

A number of the minor and some of the major varieties of goldfish now grown in America and Europe and called "Japanese" are unknown to the Japanese breeders, and were either of Chinese origin or were produced under their new occidental environment, either with or without Japanese stock. While many ephemeral freaks are necessarily produced in the course of the culture operations, the only varieties that are established and standard are those herein described.

Ten varieties of goldfish are now known and cultivated in Japan. Their Japanese names, which are most appropriate and distinctive, are in general preferable to the cumbersome and less expressive American names, and will be used in this work. These, with their literal equivalents in our language, are as follows:

Wakin, or Japanese goldfish.

Ryukin, or Liukiu goldfish.

Ranchu, or Dutch goldfish. Also called Maruko, or round fish, and Shishigashira, or lion-head. Oranda shishigashira, or Dutch lion-head.

Demekin, or protruding-eyed goldfish.

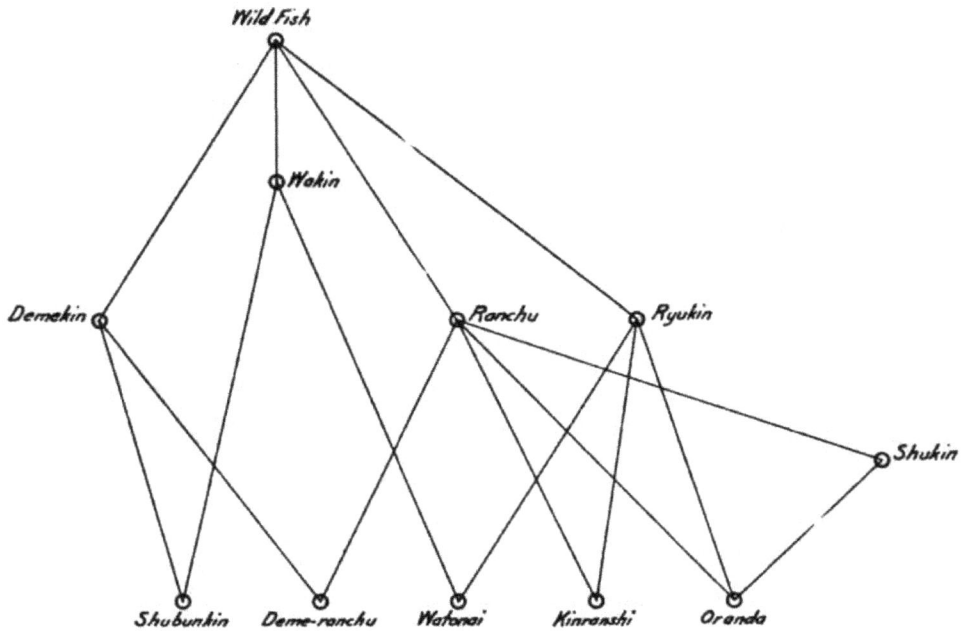

Genealogy of the goldfish varieties

Deme-ranchu, or protruding-eyed ranchu.

Watonai, or newly found variety.

Shukin, or autumn brocade goldfish.

Shubunkin, or vermilion variegated goldfish.

Kinranshi, or brocaded goldfish.

The immediate origin of the more primitive varieties can not be stated, being lost in obscurity, but the genealogy or pedigree of six of the foregoing is well known. The origin and relationships of the varieties may be represented in the form of a "family tree," as shown here.

The wakin or common goldfish

The Wakin

Japanese Goldfish. Common Goldfish

The name wakin, meaning Japanese goldfish, is applied to the simplest and most primitive cultivated variety, from which all the others have been directly or indirectly produced. As it was introduced from China, it can not properly be called Japanese. but it had been so long known and cultivated in Japan that the people of that country were doubtless justified in so designating it, especially when it became necessary to distinguish it from breeds or varieties introduced from Korea and the Liukiu or Ryukyu Islands and named after those places.

The wakin as known today in Japan. China, Europe, and America is doubtless quite similar if not identical in all essential respects to the earliest examples brought from China.

In form the wakin is moderately elongate and compressed. and the shape and size of its fins differ but little from the wild species. The caudal fin, however, may be considerably modified from the simple vertical type. The scales are large and their edges are prominent. The color is most variable. ranging from pure black to pure white or silvery, with uniform gray, brown, olive, vermilion, orange, golden, and yellow as intermediate colors, which are often variegated with black or white. The typical wakin in Japan is bright red, often with larger or smaller areas of pearly or silvery white.

A remarkable feature of the Japanese goldfish that does not occur in nature in any species of fishes and is not known to exist in any other cultivated fishes is that in many of the more highly cultivated forms the caudal and anal fins, instead of being single or unpaired, are double or paired. The caudal is the fin most subject to variation, and even in the wakin it begins to exhibit the possibility for that wonderful development met with in the more highly cultivated breeds. Three types of caudal fin may be recognized: (1) the single, unpaired, vertical form such as occurs in the wild fish and becomes more elongated and forked under cultivation; (2) the divided or paired type with the two parts united above, hence with three lobes (one medium, two lateral); (3) the divided or paired form with the two parts not united above, hence with four lobes that are more or less horizontal when spread. The second and third types are found in the most highly cultivated individuals of the wakin variety; it may be questioned, however, whether such fish. departing so much from the simple breed, are entitled to be called by this name.

As Professor Watase has pointed out, this division of the cau-dal fin is not a mere splitting of the superficial parts, but depends on an actual bilateral separation of the deep-seated bony ele-ments from which the fin arises. Professor Ryder has expressed

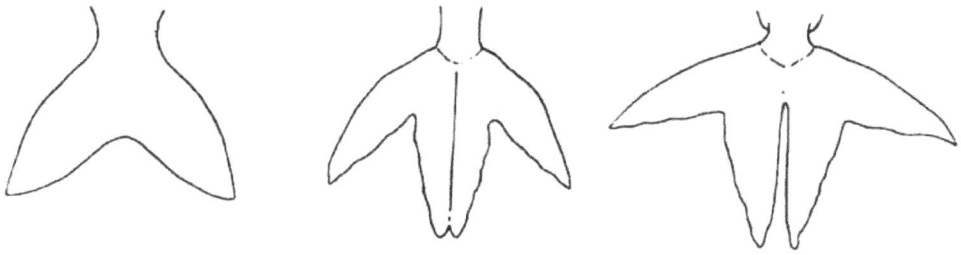

Three types of caudal fin in goldfish

the view that the double-tailed goldfish were produced originally by the orientals shaking or otherwise disturbing the eggs at the period of development when the blastoderm had spread over about a third of the yolk. This treatment of the eggs of other fishes is known to result in various forms of double monsters— double heads, partly double bodies, double tails, etc.—most of which necessarily die early. In the case of goldfish presumably produced in this way, those with double tails were most likely to reach maturity because of the least vital parts involved. "These being selected and bred," to quote Ryder, "would in all probability hand onward the tendency to reproduce the double tail, a tendency which could become very fixed and characteristic if judicious selection were maintained by interested fanciers and breeders."

The wakin is the largest of the goldfishes. Its normal length is 6 to 10 inches, and it exceptionally reaches 16 inches. It is also the hardiest, the easiest to breed and transport, and the most extensively cultivated.

When the wakin escapes from cultivation and becomes established in open waters, it reverts after a few generations to the color and form of the original wild fish, all the highly colored individuals disappearing. This has been well illustrated in the Potomac River, where the escape of cultivated fish from the government ponds in Washington has resulted in stocking the river with goldfish that are not recognized as such by fishermen and fish dealers, and are sold in the markets under the very inappropriate name of "sand perch."

The Ryukin

Liukiu Goldfish. Nagasaki Goldfish. Fringetail Goldfish

Under the name ryukin the Japanese recognize a variety that has long been cultivated and that probably was bred from ancestors similar to the wakin but became separated from the wakin stem at a very early period in the history of goldfish culture. It has been contended by some persons that the ryukin is a cross between the wakin and the ranchu, but this does not seem likely. The term ryukin is derived from Ryukyu, the Japanese rendering of the Chinese Liukiu or Loochoo, the name of the extensive group of islands lying between Formosa and the mainland of Japan; and doubtless indicates the origin of this variety or at least the route by which it entered Japan.

The ryukin or fringetail goldfish

The characteristic features of this variety are the greatly short-ened body, the rounded and bulging abdomen, and the long, flowing fins. The back is elevated, the head rather pointed in profile but broad when viewed from above, the lateral line makes a marked compound curve, and the shortening of the body in its long axis, results in strong curvature of the spine that verges on the anal fin is partly concealed by the caudal. The particular point to which this variety is bred, deformity; but this is amply compensated for by the beauty of fins and colors. The caudal exhibits .the most striking development. In the more highly cultivated fishes it is as long as the body or even longer; it is either united or split in the median line, and its delicate folds are so ample that they would completely cover the body if properly applied. The depth of the fork equals half or more than half the total length of the fin. The anal fin is either single or double, and

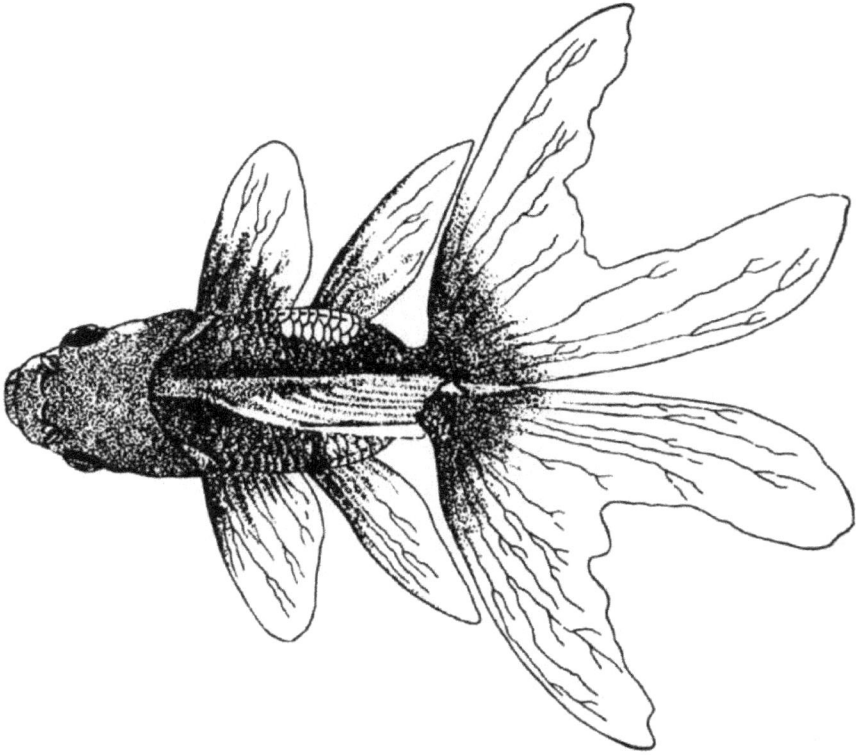

Dorsal view of a highly-cultivated ryukin

its base is nearly vertical and beneath the two parts of the caudal, while its pointed extremity may extend beyond the middle of the fork of the caudal. The high dorsal fin extends beyond the base of the caudal, and the pectoral and ventral fins far overlap the ventral and anal respectively.

The colors of the ryukin are most attractive. A unicolored fish is rarely seen, but a single color may largely predominate. The color that is practically always present is vermilion, which occurs on body, head, and fins, and is often mottled with white. A golden reflection overlies the red on the body and head, and sometimes extend on its fins. The most highly esteemed specimens are those with variegated back and sides.

The ryukin is relatively small. The maximum length from mouth to tip of tail probably never exceeds 7 1/2, or 8 inches, of which about half represents the caudal fin.

A lot of particularly fine examples of this variety presented to the United States Bureau of Fisheries by the Onuma Fish Culture Association of Iburaki, Japan, was first exhibited at the Jamestown Exposition and then in Washington. These fish have the caudal fin divided to its base and longer than the body, the anal fin long and double, and brilliant coloration.

In repose, this variety assumes either a horizontal or slightly oblique position with the head inclined upward or downward; the tail fin is pendulous and hangs in graceful folds like a soft veil; and the dorsal fin becomes folded on itself. When actively swimming the tail and caudal fin are rapidly moved from side to side in a few spasmodic efforts, but when in gentle motion the large pectoral fins are the propelling agents, and the beautiful caudal fin spreads out passively in a horizontal direction, the two halves well separated. When feeding on the bottom, the fish assumes a vertical position, and when resting on a horizontal surface the most elevated part of the body is the posterior extremity of the caudal peduncle.

A full-grown ryukin of select breed is one of the most beautiful of fishes, and would seem to satisfy all the requirements of the ordinary fancier. Such a fish, with its long, flowing, graceful fins, slowly swimming in quiet dignity, has been likened by Professor Mitsukuri to the Japanese court ladies of olden times, dressed in long robes and walking with sedate grace and dignity.

The ryukin is very extensively grown, and is exported to America and Europe in large numbers. It is a hardy variety, stands transportation well, and is altogether one of the most desirable forms for cultivation in America and one of the most attractive fishes for ponds and aquaria.

The Ranchu

Maruko. Shishigashira. Korean Goldfish

Literally translated, the name ranchu means Holland worm. The significance is not apparent, except in the fact that in early times

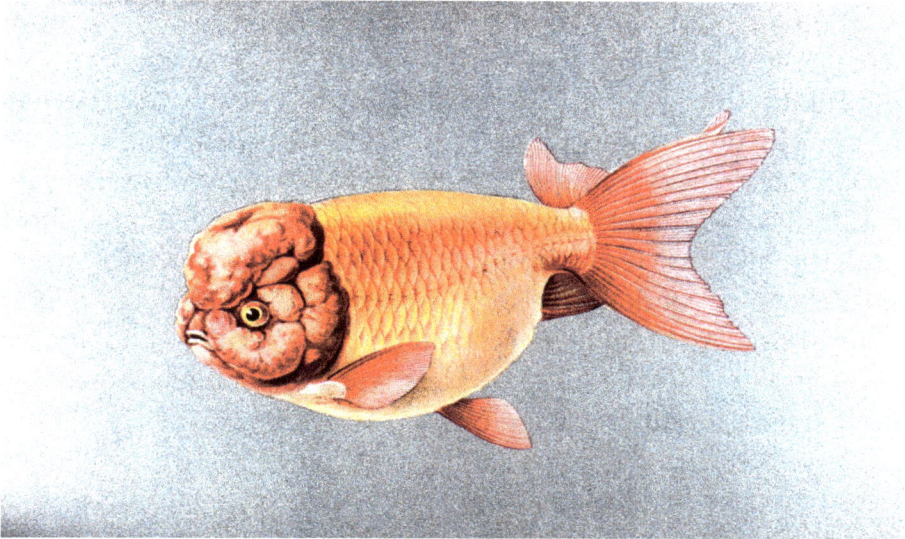

The ranchu or maruko, Korean goldfish

any new or strange animal or plant or production was regarded as of Dutch origin and named accordingly. Another and a later name in very general use is maruko, meaning round fish. This variety is called also Chosen, or Korean, goldfish, in allusion to a possible origin or a possible route from China to Japan. There is usually a peculiar growth or wart-like protuberance on the head, and this gives rise to other names—shishigashira and shishigashira ranchu, lion-head and lion-head ranchu.

The wide departure of the ranchu from the original form suggests that the parent stock must have been a very early offshoot of the wakin stern, probably earlier than the ryukin and the demekin, which are the other more primitive types now grown in Japan.

The ranchu is easily recognized by its short, rounded body, broad head, short caudal fin, and deficient dorsal fin. As the length, depth, and breadth of the body are about equal, and as the back and belly are broad and rounded and the caudal peduncle very short, the form of the body proper is almost globular and a median cross section would be a nearly perfect circle. The head is short and as broad as deep, and the snout is broad, short, and rounded. The back is nearly straight or only slightly elevated and shows no trace of a fin. The caudal is short and three-lobed, with the lobes rounded and the two lateral ones having a tendency to spread horizontally. The pectoral and ventral fins are small and present no peculiarities. The anal is short and double. A curious sickle-shaped appendage sometimes appears on the dorsal edge of the caudal fin.

Up to the time the fish is two or three years old, the head does not show any peculiar features, but it then begins to develop a mass having the appearance of a warty tumor. In some specimens the warts are of uniform size and very regular distribution, in others they are irregular in size and shape. The warts are soft to the touch, and represent simply the enlargement of the normal papillae of the skin of the head; and so far as known the mass does not take on any malignant character.

The amount of surface covered by the growth varies, and this, together with differences in the warts themselves, gives rise to several subvarieties. In the lion-head proper the entire head except the lower jaw is covered with large red, pink, or white warts, and the head and snout are thus greatly broadened. In

Types of the ranchu or maruko variety of Japanese goldfish; from a water-color painting from life, made for the author in Tokyo by K. Ito

the form known as the tokin, or capped or hooded goldfish, there is on top of the head a mass of warts projecting one-half to three-fifths of an inch above the surface and sharply defined all around. The warty growth is sometimes entirely white, and may contrast strongly with the colors of adjacent parts. Fish thus colored are called hiragashira, or white-heads. As white warts are nearly always smaller than red warts and the growth is thus less prominent. These fish are known also as shiragashira, or flat-heads.

The color of the ranchu is quite variable. Originally the color seems to have been more or less uniform, and at present the most popular fish are those with a uniformly golden or red body and a bright red head ; about twenty years ago, however, examples with variegated backs began to be produced, and such are now cultivated in large numbers. A striking color phase sometimes met with is a white fish with individual, regularly distributed scales of the back and sides bright red, and with the head pink. An otherwise white fish may have bright red fins and red head.

A full-grown ranchu is 6 inches long, including the caudal fin. The largest size attained appears to be about 7 1/2 inches.

The ranchu is a clumsy fish, with swimming powers reduced to a minimum. The absence of dorsal fin combined with the nearly globular body has resulted in a loss of ability to maintain a normal position, and in consequence the fish often swims upside down or vertically with the head downward. The variety is weak, delicate, and difficult to keep; and although very extensively grown in Japan has rarely been exported to foreign countries. Limited numbers have been brought to the United States from Europe or directly from Japan, but the fish is known to very few people in this country. Its cultivation should be more generally undertaken both on account of itself and because of the opportunities afforded for producing new forms by crossing with other varieties.

The Oranda Shishigashira

Dutch Lion-head

The oranda shishigashira, or Dutch lion-head, did not originate in Holland, but was so-called according to Japanese custom because it was a new or strange production. It was first bred at Koriyama or Osaka about 1840, and was produced by crossing the ranchu and the ryukin. It possesses the peculiar cephalic excrescences of the former and the flowing drapery of the latter, and is perhaps the most striking type of goldfish, representing the highest development along the two lines indicated.

The body is less elevated than in the ryukin and more elongate than in the ranchu. As adult age is neared, a warty mass develops on the head; this excrescence varies in extent, sometimes covering the entire top and sides of head, sometimes forming a cap from the eyes backward. In some specimens the normally broad head is still further broadened by the lateral growth, and the eyes are invisible from above. All of the fins are large. The dorsal base is long and occupies two-thirds of the length of the back, sometimes arising near the head and sometimes much posterior thereto. The caudal is three-lobed, four-lobed, or bag-like, and may equal or even exceed the entire length of the body of the fish.

The oranda shishigashira or Dutch lion-head

In its graceful folds and delicate texture the caudal resembles that of the ryukin. The long, double anal surrounds the terminal vent.

In color the oranda is either plain or variegated. Up to twenty years ago fish with variegated backs were not known, but since then red-and-white fish have been common. The plain red fish never show the rich golden iridescence of the ranchu. Some fish are uniformly velvety black with golden reflections below; some are red with the abdomen canary yellow instead of white; and various other colors are met with. The warty mass may be white, pink, vermilion, orange-red, black, or variegated. The fins are usually bright red, with more or less white on the caudal. A strikingly beautiful large male oranda seen by the author in Osaka had a red head, a yellow-golden body, a black back, and black fins.

This variety when originally produced was only 2 to 3 inches long, but now it is one of the largest forms cultivated. The

Dutch lion-head, from a water-color painting from life, made for the author in Tokyo by K. Ito

ordinary length of mature fish is 4 to 5 inches, with the caudal fin as much longer. The largest specimens have the body and caudal each over 6 inches long, and weigh nearly 20 ounces.

The oranda shishigashira is extensively bred in Tokyo, Osaka, Koriyama, and other places, and is one of the favorite varieties, combining the beautiful and the curious in a striking degree. In hardiness it is between the ranchu and the ryukin.

In a number of places in Japan a rather strongly marked sub-variety of this form has been developed, the peculiarities being a short tail and a brassy color; it is called the shishi, or lion, and is perhaps entitled to be considered a distinct variety.

The Demekin

Popeye Goldfish. Telescope-fish

The name demekin is given by the Japanese to a goldfish well known in Europe and America as the telescope-fish. The Japanese term, which signifies popeye goldfish, is much to be preferred; dente means "protruding eyes." "Telescope" is a singularly in-appropriate name, because the eyes are not telescopic, i.e., long-sighted, but are extremely myopic, or short-sighted. The Chinese call this variety the dragon-eyes.

Although this variety is almost always called Japanese in for-eign writings, as a matter of fact it was developed in China and was unknown in Japan until the close of the war with China (1894–5). The Japanese have, however, improved on the original importation. The variety doubtless came into existence at a com-paratively early date, and has undergone much modification of form and color in China.

The special feature of this variety is the lateral protrusion of the large eyeballs. The extent of the protrusion differs in different types or individuals, in some being very slight, in others strongly marked. The age of the fish modifies the condition; when first hatched and for about a year thereafter, the eyes are normal in size and position, but in the further course of growth the

The demekin or telescope-fish

protrusion gradually develops. Sometimes even in full-grown fish only one eye protrudes, the other being normal.

The body is rather short, the back is elevated and the ventral outline much decurved; the head is broad, and the snout is rounded and very short. As originally introduced from China, the demekin had a short caudal fin, but the Japanese have given it a long, flowing caudal. and have improved the fish in several other minor particulars. The anal is double, and it and all the other fins are long.

The demekin rarely shows a brilliant coloration. The usual colors are a uniform black, or a pale red or pale reddish-yellow with small black spots or irregular black areas; in the variegated form the fins may be reddish, blackish, or blackish with pale red or orange base. Sometimes, especially in fish of pure Chinese breed, there may be three or four irregularly distributed or mottled colors in a single fish.

This variety is rather small. A fish with body 5 inches long is considered very large, and the average is much less. The caudal fin is shorter than the body proper.

This curious and interesting variety is now extensively grown in parts of Japan, and it or the original Chinese form is common in America and Europe. The fish has very defective sight and is unable to adapt itself to the protruding eyeball, for in the adult stage it is very likely to injure the eyes by swimming against hard objects, and so becomes blind. The fish is solitary in its habits, and does not swim with its fellows, in this respect differing from all the goldfish long cultivated in Japan.

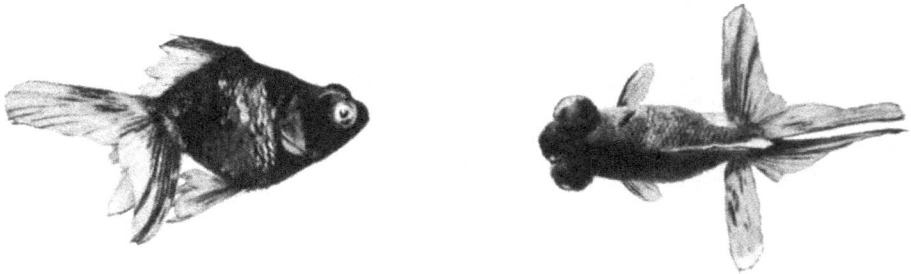

The Deme-Ranchu

Popeye Ranchu. Telescope-fish. Celestial

The Japanese have bestowed the name deme-ranchu on a variety whose acquaintance they have very recently made combining characters of the demekin and the ranchu. Americans became familiar with this fish much earlier than the Japanese, and called it the celestial, in allusion to the peculiar direction of the eyes.

Professor Mitsukuri suggested the name "astronomical telescope-fish" for the same reason.

When a person sees this variety for the first time, he is likely at once to assume a Chinese origin from its grotesque appearance. The fish did in fact originate in China, and was unknown in Japan until 1901 or 1902, although long before that time it was often ascribed to Japan in western countries. Demekin and ranchu are evident in its construction, although the former factor must have been of a somewhat different type from that known in Japan.

The general shape is like the ranchu, the body being shortened, the vertebral column strongly curved upward, the back and head very broad, and the abdomen distended laterally, so that the globular form is approached. The eyes are very large, and in highly developed, full-grown fishes their diameter is more than half the length of the head. The "telescopic" feature varies from a moderate bulging to an extraordinary degree of protrusion that may exceed the diameter of the eye. Along with this elongation of the eyeball there is a tendency to turn upward, and in the typical deme-ranchu the eyes no longer point horizontally, but vertically, having changed their direction 90 degrees, and the pupils look straight toward the sky. It is to such fish that the name "celestial" applies.

The dorsal fin is absent. The caudal is long, widely spreading, and with the lower lobes extending at right angles to the long axis of the body; its length may exceed that of the body.

The coloration is similar to that of the demekin: there may be a uniform black or pale orange, or both of these colors may be present in varying proportions.

The dislocation of the eyeballs in this variety goes so far as to produce a genuine monstrosity. The fish has very feeble swimming powers and seldom exerts them, but remains solitary at the bottom of the aquarium or pond resting on its abdomen.

The deme-ranchu or popeye ranchu

The vitality is low, the ability to reproduce is impaired, and spawning occurs only rarely. For all these reasons the culture of the deme-ranchu is not popular or extensive.

The Watonai

Fringetail Wakin

The euphonious Japanese name watonai, according to Professor Matsubara, means "a variety hitherto found neither in Japan nor China." English names that may be used are Japanese fringe-tail and fringetail wakin.

The watonai was produced naturally in a pond containing brood specimens of the wakin, ryukin, ranchu, and oranda shishigashira, and represent; a hybridization of the two first-named varieties. It came into existence in Tokyo about 1880, and was first exhibited at a fisheries exhibition held in Tokyo in 1883.

The watonai or fringetail wakin

In general form this fish is similar to the wakin, but the body is shorter, thicker, and rather deeper, and all the fins are larger. The caudal is shaped as in the ryukin, and is nearly as long as the body. The colors are usually variegated red and white.

This fish, whose parentage is so apparent, combines the hardiness of the wakin, the long, graceful fins of the ryukin, and the rich coloring of both. Its size equals or exceeds that of the ryukin, but does not reach that of the largest wakin.

The Shukin

Autumn Goldfish. Longtail Ranchu

The name shukin was given by Professor Matsubara to a variety produced at Tokyo in 1897, and apparently independently at Osaka about the same time, by the crossing of the ranchu and the oranda shishigashira. The name means literally "autumn

The shukin or longtail ranchu

brocade," and was applied in allusion to its bright red coloration suggestive of the beautiful autumnal foliage of the Japanese maples. This variety was not perfected immediately but required several generations of selected breeding of the progeny of three-year old fish of the two varieties mentioned. The form is now well established, but has a tendency to reassert characters of the oranda which were sought to be eliminated.

The shape of the body and head is typically ranchuan, although the lateral swelling of the abdomen is not excessive. The head shows, in variable degree, the warty growth on the head that is characteristic of both parents. All of the fins are longer than in the ranchu. The four-lobed caudal is as long as or longer than the fish proper, and has the peculiar texture that permits the most graceful drooping and waving. The anal is double and long, and the ventrals extend far beyond the base of the caudal. The dorsal fin is absent, hut in the first generation of the cross this fin appears in over ninety per cent of the young, usually in the form of one to three rudimentary rays.

The shukin of select breed has a bright golden or red body, a red head with red excrescences. and red-and-white fins; but variations in color are numerous.

A size of 9 to 10 inches is attained, the caudal fin constituting about half the length. A five-year-old fish examined by the author in Tokyo had a body 4 1/2 inches long and a caudal fin somewhat longer.

This is one of the most beautiful and attractive varieties, and will well repay efforts at cultivation. Of the three recently perfected varieties, it is the most popular among the Japanese. Owing to its greater development of fins, its swimming powers are superior to those of the ranchu.

The Shubunkin

Speckled Goldfish. Spotted Goldfish. Vermilion Variegated Goldfish

According to its sponsor, Professor Matsubara, the name shubunkin means "vermilion red dappled with different hues." As distinctive English designations, the names speckled goldfish, spotted goldfish, and vermilion variegated goldfish may be employed.

This is a large, graceful form, showing combinations of color not found in any other Japanese varieties. It came into existence in 1900, and is the outcome of the breeder's efforts to produce a fish that shows the multivariegated speckles or mottlings sometimes seen in the Chinese demekin but not in any of the older varieties grown in Japan. The hybrid was effected. according to Professor Matsubara, by the mating of the demekin and the wakin, of which an equal number of each sex and of each variety was selected for the purpose, the demekin having black dapples

The shubunkin or speckled goldfish

on a vermilion or purplish body, and the wakin being variegated with red. black, bluish, and white. The results of this cross were interesting as showing the possibilities of further experiments of this kind. Some of the young had the form of the wild goldfish and the peculiar markings of the demekin; some resembled the wakin; some had the form of the demekin. About twenty per cent of the progeny were of the special type sought to be produced. This has been regularly bred, and has given rise to some new and interesting color phases.

In the standard fish of this variety the body is rather long and compressed. the dorsal and ventral outlines are decidedly curved, the caudal peduncle is very distinct, and the scales are less conspicuous than in other varieties. The dorsal fin is elevated and wavy; the pectorals. ventrals, and anal are moderately elongated, the last being simple; and the caudal is bilobed and deeply forked, and three-fifths to two-thirds the length of the body.

The kinranshi or brocaded goldfish

The shubunkin normally has a peculiar mottled coloration, with small, irregularly distributed black spots on body and fins as a characteristic feature. The mottling in one individual may comprise vermilion, together with black, white, bluish, purple, or other colors. The color is often bright red, spotted with black; and occasionally a fish is produced that is uniformly purple—something quite unknown in the parent stock on either side.

The Kinranshi

Brocaded Goldfish

This latest addition to the Japanese varieties of goldfish was called kinranshi by Professor Matsubara, the literal meaning of the name being "goldfish with brocaded figures." The variety was produced a few years ago by Mr. Akiyama Kichigoro, a celebrated goldfish culturist of Tokyo. The desire being to develop a new variety lacking the dorsal fin, 20 selected male and female ryukins were crossed with the same number of selected ranchus, with the result that in one-third of the progeny the dorsal was entirely absent, while in the others the dorsal was normal or was represented by spines or protuberances.

This variety, as now established after five or six generations, has an elongate and rather thick body, gently arched back, and small fins, the caudal being double. The colors are showy, consisting of red, black, and white in varying proportions.

3 Goldfish Breeding Establishments, and Their General Equipment and Management

Goldfish Farms

The cultivation of goldfish in Japan is conducted in open-air ponds, not in aquaria, troughs, or tubs within doors as is often the case with goldfish fanciers in America. The goldfish farms are necessarily much alike and are conducted in the same general way, the principal differences being those dependent on the magnitude of the operations. The number and size of the ponds vary considerably; some breeders have only a few ponds of small area, other s have numerous ponds with a very large aggregate area. There are, however, differences in the pond systems occasioned by local conditions, by individual or regional practices, and by the particular varieties to which most attention is given; and different methods of culture are required for the different kinds of fish.

The proprietors of the farms have their residence in close prox-
imity to the ponds, often surrounded by them; and they carry
on a large part of the practical work in rooms in which they and
their families live. There is no such thing as a goldfish hatchery
in the sense of a special building devoted to the purpose.

To inspect any of the more extensive goldfish farms in Tokyo
or Koriyama is a most interesting and delightful experience,
whether a person be a goldfish fancier or only a casual visitor.
The sight afforded by the fishes of different breeds and sizes,
with their brilliant coloration, graceful appendages, and peculiar
movements, can hardly be rivalled at any other fish-cultural
establishments in the world. Additional attractions usually to be
seen are little ponds containing tortoises, red and variegated carp,
and various other water creatures cultivated for use or ornament.

A very old goldfish breeding station in the outskirts of Tokyo
inspected by the author has a pond area of 44,000 square feet
and an annual output of about 500,000 goldfish, in addition to
which golden carp and common carp are grown. Ten persons are
employed in the various branches of the work, and six varieties
of fish are handled: wakin, ryukin, ranchu, oranda shishigashira,
demekin, and shukin. The principal goldfish establishment at
Koriyama has 50 large ponds and numerous small ones with an
area of perhaps 150,000 square feet, including extensive ponds
reserved for the growing of food for the young and old fish.
Five varieties are here regularly grown, and the annual crop
sometimes exceeds 600,000 fish.

The accompanying illustration shows a part of a typical Tokyo
goldfish farm. In the foreground are shallow breeding ponds in
which the water has been drawn down. The several flat dishes
suspended from bamboo poles stuck in the muddy bottom are
the receptacles on which food for the goldfish is placed. Further
back is a cluster of small, very shallow cement ponds or basins

View in a Tokyo goldfish establishment

among which a man is standing; over some of these basins the wire-gauze covering has been tilted back. At the extreme left an employee is drawing water from a shallow well with a sweep. In the small house on the right fish food is prepared. The larger house on the left is the home of the proprietor and his family, and is the business headquarters of the plant.

General Principles of Breeding

The remarkable results that have been achieved by the Japanese in producing variations in the form and color of gold-fish have depended on no secret or mysterious processes and no mechanical devices or appliances, but have been due to an intelligent application of natural laws. Professor Mitsukuri, of the Imperial University of Tokyo, has noted that the Japanese goldfish culturists well understand the principle of "breeding to a point," and although they are usually without much education and have acquired all of their knowledge from practical experience, they often discuss evolutionary matters in a way that suggests acquaintance with the Darwinian theory of the origin of species. Some of the ideas current in America as to the ways in which the varieties have been produced are quite absurd and cause much amusement to the Japanese.

In brief, the Japanese breeders have attained their success by adhering as strictly as possible to nature in feeding, rearing, and otherwise caring for their fish; by eliminating the unfit; and by providing a superior brood stock and definitely selecting the fish that are to mate each year. The resulting rich harvest, with its beauty of form and color, is a necessary consequence, and bears testimony to the combination of patience, skill, and intelligence in the character of the Japanese that enables them to accomplish so much in all their pursuits.

The only exception to the employment of purely natural methods in Japanese goldfish culture is that at Koriyama the practice has existed from very early times of artificially making designs on the backs of the fish. This is done by the use of dilute hydrochloric or muriatic acid, and the process consists in a decolorization which leaves the treated parts white. This destruction of the pigment in the skin is possible only over the scaly body, and cannot be safely brought about on the unsealed parts—the heads and fins. The operation is best performed in August or September, and the fish should be in the highest physical condition by having had an abundance of fattening food for some time before. By the use of a brush, glass rod, or stick, the acid is judiciously applied after the skin has been wiped dry. In this way flowers, figures, letters, etc., are produced; but the results are not very pleasing to the esthetic taste, and the practice is quite uncommon and is not to be commended.

The methods herein described are such as are followed at Koriyama and Tokyo and by the principal culturists at those places; and the information given is based on the personal observations of the writer. Use has been made of additional material contained in several excellent papers by Japanese scholars.

The Ponds and the Water Supply

In growing goldfish for profit it is necessary to have ample pond area. The extent of the business will of course depend largely on the amount of water in which the fish may be grown; and the season's success may often be affected by the number of available ponds into which young and adult fish may, in emergencies, be transferred.

There are two general classes of goldfish ponds, large or mud ponds and small or cement ponds; to these, in the Koriyama district, are to be added special food ponds.

The mud ponds are so called because they have a soft mud bottom, and are usually roily in consequence. They are rectangular, and their size depends on local conditions, individual tastes, number of fish to be held, and particular variety of fish to be cultivated. In Koriyama, one extensive breeder has ponds as large as 50 by 100 feet, while other culturists prefer ponds of smaller and more convenient size, say 18 by 50 to 60 feet. The depth of these ponds never exceeds 3 feet and often is only 1.5 to 2 feet. This shallowness is an important feature, ensuring efficient oxygenation and lighting, and would doubtless be carried still further if it did not expose the fish to injury from too strong sun's rays and sudden atmospheric changes. The sides of the ponds are formed of upright boards sunken into the bottom; and the ponds are separated by gravelly or sandy walks bordered with grass, flowers, or other vegetation. By means of a gate or pipe the ponds may be drained as needed for cleansing and other purposes.

The mud ponds are for brood fish, for grown fish intended for sale, and for fish in course of rearing. They are kept constantly stirred by the swimming and feeding movements of the fish.

In modern goldfish culture small cement ponds are quite as indispensable as the larger mud-bottom ponds. Their size depends on personal preference and the purposes for which intended. The dimensions may be as small as 3 by 3 feet or as large as 12 by 12 feet, with all intermediate sizes. with a depth of 6 inches. In Koriyama, for the accommodation of the large oranda variety, ponds are 18 to 20 feet long and 5 feet wide, with a depth of 8 to 10 inches. Such ponds are usually and most conveniently arranged in series of 6 to 12, sometimes separated by narrow walks, sometimes only a few inches apart. Each set of ponds, or basins, as they should perhaps be called, is supplied with water through a common open trough or flume; and each basin has a watergate in the middle of the side next the flume. The outlet pipe occupies the center of a rounded depression near the opposite end of the basin; this concavity is 9 to 18 inches wide and 2 to 4 inches deep, and is intended to receive the fish and prevent them from struggling as the water is drawn off. These basins are usually provided with covers or awnings so as to regulate the amount of light and to afford protection from enemies and elements.

The cement basins are used for the retention of brood fish immediately prior to and during spawning, for the hatching of the eggs, for the rearing of the young, for the holding of fish awaiting sale and shipment, and for various other purposes.

The water supply of goldfish establishments is generally far from profuse, and often of a character that would seem to be questionable; nevertheless it serves the purpose admirably. The ponds are flooded with water from various sources, but only rarely are they supplied by gravity from a running stream. Shallow wells provided with pumps or buckets may be found at nearly every station; stream water when available is transferred by pump, treadmill, or buckets; ditches are often drained into the ponds; and rainwater is generally utilized. The ponds being quite sluggish and subject to infrequent change or renewal, algae often grow rankly and give the water a distinct greenish color. There are no rooted and surface flowering plants in the ponds, such being rigidly excluded. In some instances the ponds receive the discharge of gutters of the town or city, such water being considered desirable because it contains a large amount of organic matter whose decomposition favors the ultimate growth of fish food.

In order to guard against the development of poisonous gases and other deleterious substances, and also to eliminate enemies that may have entered, the mud ponds are drawn down at least twice a year (spring and late autumn), extraneous substances raked out, and the bottom exposed to the air for 4 to 6 days. More frequent draining is desirable if practicable.

4 The Parent Fish, the Egg-laying, and the Hatching

The Care, Selection, and Mating of the Brood Fish

FISH from which it is expected to get eggs at the next spawning time are given special attention in autumn, and are then provided with an abundance of suitable food so that they may begin the winter in a robust state and emerge therefrom in the best possible condition. Another reason for promoting the physical well being of the fish particularly in the autumn preceding spawning is that they may then develop their colors and shape most fully and give the culturist the best information as to the possibilities of his brood stock.

The spawning season extends from the latter part of March to the middle of June, but April and May are the chief months. The time when particular fish lay their eggs may be controlled to a certain extent. Fish that are given sufficient food and retained in stagnant water will have their spawning retarded or altogether inhibited; while fish that are exhibiting the symptoms of approaching spawning may be made to deposit their eggs within one or two days if they are given plenty of food and have the water in their pond frequently changed or if they are transferred to another pond.

At the approach of the spawning time the fishes' colors become brighter, the abdomen in the female begins to enlarge owing to the growth of the ovaries, and there appear on the head of the males peculiar excrescences ("pearl organs") that may be too small easily to be seen but are readily detected by touch. The fish crowd together in the ponds and make much commotion as they splash and jostle. They eventually separate into pairs, or rather each male attaches himself to a female, pursuing her, swimming around her, and rubbing her abdomen with his roughened snout and opercles. Sometimes two or three males will follow a female. As the time for spawning comes nearer, the attention of the males becomes more assiduous and the ripening and loosening of the eggs are doubtless facilitated by their actions.

Goldfish begin to breed when two years old and continue to spawn for six or seven years or even longer, but the best brood fish are those that are three, four or five years old. The fish three and four years old are the most satisfactory. After the fifth year the spawning capacity rapidly diminishes, and fish so old, having served for breeding purposes, are usually sold and make useful aquarium objects for many years thereafter. The normal age attained by the more hardy varieties is sixteen or seventeen years.

The fish of suitable age for breeding purposes are subjected to careful and critical examination, and a selection is made of those whose mating is most likely to produce the qualities most desired in the offspring. In addition to physical vigor, the general form of body, character of fins, and pattern of coloration are duly considered, with reference to the special fashions in vogue in the community and the requirements of the trade.

Removing dust, dirt, bubbles, etc., from the surface of a concrete spawning pond into which a new supply of water has just been run

The Spawning Ponds and their Preparation

As the ponds in which the brood fish are kept do not contain materials suitable for the reception of eggs, it is necessary either to insert such substances or to transfer the fish to ponds that have been prepared for the purpose. The latter course is preferable for various reasons.

In Tokyo the favorite articles for spawning beds are living water plants, particularly the milfoil (*Myriophyllum verticillatum*) and the hornwort (*Ceratophyllum demersum*), but in Koriyama preference appears to be given to roots of the willow (*Salix*).

Several weeks before egg-laying time the fine, matted roots of the willow are collected in large quantities, thoroughly washed, then boiled to sterilize them, and finally dried.

The usual procedure, when spawning is imminent, as shown by the behavior of the fish and the temperature of the water, is to transfer the fish to the spawning pond, the sexes being about equally represented and the number depending on the size of the pond. The water plants or the bundles of willow roots are placed in this pond, and on these the eggs are soon deposited.

The common practice at Tokyo, in the case of the ranchu, for example, is to hold the ripening fish in concrete basins. A small cultivator might have only three pairs of fish in a pond, while an extensive cultivator might have twenty-five pairs. It is considered unnecessary to change the water, and the principal attention the fish require is to be amply fed for about ten days before spawning, the preferred food at this time being worms and mosquito larvae. As the experienced cultivator can usually tell when the eggs will be laid, the water in the spawning pond is renewed the previous day, the milfoil is introduced, and the brood fish are removed thereto. For three or four pairs of fish a pond or basin with an area of ten or twelve feet is sufficiently large, while for twenty-five pairs the pond should be forty to forty-five feet in area. Then, should the atmospheric conditions be suitable—a rise in temperature or a warm rain—the fish will spawn the next morning.

When cultivated after the Tokyo method, the brood fish of the ryukin variety are kept in mud ponds and are permitted to spawn in the same pond in which they have wintered. The ripening and deposition of their eggs are encouraged by giving a plentiful supply of food, and the growth of natural food within the pond is facilitated by the use of fertilizer, as hereinafter noted. When the temperature of the pond has risen to 60°F., as usually happens about the first of April or earlier, the water is renewed and the material for the spawning beds is inserted. As many as 4·00 to 500 pairs of fish three years old are allowed to spawn in one pond, the proper space for each 100 pairs being about 400 square feet.

In the breeding operations at Koriyama, the parent fish are generally kept in large mud ponds and deposit their spawn there, and as the eggs are laid they are transferred to concrete ponds for hatching.

When it is the desire to produce orandas of the largest size, parent fish six years old in sound physical condition and with good form are put into a special roomy pond, allowing about four square feet per pair, and are given an abundance of suitable food; and to carry the cultivation for size still further, brood fish seven years old are selected and are allowed ten to twelve square feet per pair.

The Eggs, their Care and Development

As is the case with the vast majority of fishes, the eggs of the goldfish are fertilized after deposition. While artificial propagation is doubtless feasible, it is entirely unnecessary and is not practiced in Japan or elsewhere because under natural conditions fertilization is ordinarily most perfect.

When the eggs have become mature within the ovaries and the female is fully ripe, the extrusion of the eggs is accomplished by a series of spasmodic muscular efforts. At the same time or immediately thereafter, the attendant male emits the milt that contains the fertilizing cells, which are disseminated throughout the adjacent water and come in contact with the eggs.

Goldfish eggs are slightly heavier than water and are not adherent to one another, so that when they are expelled they

Goldfish spawning on willow root in a mud pond

settle on the roots or water plants that constitute the spawning bed and cover them more or less evenly. Their surface being sticky, the eggs have a tendency to remain where they first settle, and as the mucilaginous material quickly hardens in water the eggs are securely held in a position most favorable for thorough aeration while hatching.

Unlike many other fishes, the goldfish exercise no care or solicitude for their eggs when they have once been laid, but on the contrary promptly devour them if permitted to do so. It is therefore necessary to remove either the parents or the eggs to another pond. When the eggs are to be transferred from one pond to another for hatching, the bundles of roots or the plants containing them are gently washed in clear running water, and carefully placed in the hatching ponds. In order to prevent the eggs from becoming crowded or smothered, the bundles of roots are sometimes tied on a rope at regular intervals and arranged in rows. Having regard for the accommodation of the fry during the days immediately after hatching, the proper number of eggs for a cement pond with an area of 100 square feet is 50,000 to 60,000.

The different varieties of goldfish produce about the same number of eggs when fish of the same size and age are considered. At Koriyama, the oranda will lay approximately 2,000 eggs when two years old, 25,000 eggs when three years old, and 70,000 eggs when four and five years old. The eggs in different parts of the ovaries do not ripen at the same time, and the spawning period for a given fish is thus quite prolonged. Individual fish deposit from three to tell lots of eggs at intervals of eight to ten days. The first batch of eggs is the best, the last is the worst and is likely to produce weak fry.

The goldfish egg when first deposited has a slightly wrinkled and loose outer covering, but owing to the fact that it immediately begins to absorb water it quickly assumes a perfectly spherical shape and the limiting membrane becomes smooth and tense. The average diameter of the fertilized egg is .0625 inch, and the number in a pint is about 137,500. A viable egg is transparent and colorless or slightly yellowish, but an unfertilized egg soon becomes milky and opaque.

The only attention the eggs require is to see that they are covered with water, are not becoming fungoused, and are protected from unfavorable weather conditions. Should hail threaten, a strong wind blow, or the air temperature fall suddenly, the ponds must be quickly covered with matting or screens.

As a general thing, the hatching ponds receive and require no fresh water while incubation is in progress. It is sometimes desirable, however, to effect a change of water, particularly if the temperature becomes very high.

The hatching period is comparatively short, and normally occupies eight to nine days at a water temperature of 60 to 65°F. A rising temperature may reduce the hatching time as much as one-half, but a very rapid development of the eggs is unfavorable. On the other hand a fall in the water temperature may prolong the hatching one-third to one-half, but a greatly retarded incubation is likewise unfavorable, especially in that the young are not of uniform size. A peculiarity of the goldfish egg is that the embryo covers nearly the entire circumference of the vitellus, and the yolksac is comparatively small and very granular.

5 Food, Growth, and Care of the Fish

Living Crustacean Food and its Cultivation

PROPER food in proper quantity at the proper time is of the most vital importance in successful goldfish culture. Not only do the life and growth of the young depend absolutely on it, but also the form, color, spawning capacity, and market value of the resulting adult fish. The natural and therefore the best food for goldfish at all stages of growth is minute crustaceans belonging to the sub-class Entomostraca, particularly those of the orders Copepoda, Ostracoda, and Cladocera, which are often popularly included in the general term of "water fleas," and are referred to by the Japanese as "mijinko." Among the best known and most important of these are *Cyclops, Cypris, Daphnia,* and *Polyphemus.*

These little creatures occur naturally in nearly all fresh waters, and abound in the ponds in which goldfish are reared: but under ordinary conditions the supply would soon be exhausted even though they multiply quickly. Therefore, one of the chief duties of the Japanese goldfish breeder is to devise ways and means to insure an abundance of such food. To this end special waters must be available for the collection and retention of such creatures, and special efforts must be made to encourage their growth in the goldfish ponds.

The most striking feature of goldfish culture as practiced by the Koriyama school is the great amount of effort and time devoted to the collection and artificial production of crustacean food. This subject of course receives attention at Tokyo and elsewhere, but is less characteristic and in general much less elaborately worked out than at Koriyama.

The simplest method of providing these small crustaceans is to collect them in open waters—reservoirs, ditches, ponds, streams—if the culturist has convenient access to such. The usual collecting grounds are the reservoirs for the irrigation of rice fields, in which the conditions are very favorable for the existence of these creatures. Many are produced also in the mud ponds.

The enterprising breeders, however, do not depend on the natural growth of "mijinko," but resort to artificial measures for maintaining a constant supply. This work is scarcely less noteworthy than the cultivation of the goldfish themselves, and is perhaps the most remarkable feature of this industry, for the Japanese have gone far ahead of other people in this important branch, As Professor Mitsukuri has said, "the Japanese goldfish breeders have the knack of producing these water fleas in any quantity they need at any time they like."

The essential point in the cultivation of "mijinko" is the fertilization of the pond, so that the growth of the minute animals and plants that serve as the immediate or the ultimate food of the crustaceans may be greatly stimulated. Given an abundant food supply, the little crustaceans will multiply with astonishing rapidity and soon acquire a bulk which in the aggregate is very considerable. There are several ways of fertilizing the mud ponds. Reference has already been made to the use at times of water from the gutters and ditches of the towns and villages; this water, rich in organic matter, both living and dead, is run directly into the rearing ponds. Another practice is to place in the ponds loose rolls of straw matting permeated with rice bran or the lees of soy; the fermentation that ensues finally promotes the growth of the desired crustacea. Another method is to put soy lees in a loose straw bag that is placed on the bottom of the pond, with the same results as before, the crop of water fleas being ready for harvesting by the young goldfish at the expiration of 4 or 5 weeks. Still another way of inducing quickly an abundant growth of crustaceans is to inoculate the pond therewith; a pint of these creatures placed in a suitable pond of an acre or 1,000 to 1,200 square feet will multiply so rapidly that after 3 to 5 days many thousand young goldfish may be subsisted.

For the more effective and extensive production of "mijinko," however, it is customary to proceed somewhat differently. Supposing a pond to have a area of 1,800 square feet, there will be placed in it rice bran, soy lees, or fresh horse manure to the amount of 4 or 5 bushels, the pond having previously been drained; for a recently constructed pond the quantity of fertilizer must be increased and a combination of soy lees and manure may he used. After the bottom of the pond, with the fertilizing substance spread over it, is exposed to the sun for 6 to 8 days, the water is turned on and the pond is flooded. In a few days the color of the water becomes decidedly green from the presence of unicellular algae in great abundance, and in 2 to 4 weeks the water fleas will exist in such numbers that they will support many thousand young goldfish with constantly increasing appetites. It may be necessary, however, to continue to apply fertilizer to the pond at short intervals.

The collecting of entomostraca from reservoirs and lakes for the newly-hatched fish in the cement ponds is an important part of the work and occupies the time of many people. The peculiar bag-net required for this purpose is 25 to 30 feet long and 2.5 feet wide throughout its length, the bottom being cut off square and gathered with a string; it is made of fine cotton or calico that has been treated with extract of oak bark or other astringent solution. The bamboo pole to which the bag is attached is of about the same length. One man, standing on the bank, operates the net, drawing it slowly through the water and gradually accumulating a mass of water fleas and in addition insects, plants and various other kinds of material which must be separated from the crustaceans.

As many species of entomostraca are entirely too large for the mouth and stomach of the goldfish fry, it is necessary to sort them according to size, and this procedure is required during all the stages of the fry period. The separation of the crustaceans into sizes adapted for the different sizes of the fry is accomplished by means of sieves, of which 5 sizes are used, having respectively 130 meshes, 100 meshes, 80 meshes, 60 meshes, and 20 meshes to the inch. The sieves are 10 to 15 inches in diameter, and have either wire or cloth bottoms. Besides separating the crustaceans, they serve to exclude foreign matter in the water; and the coarsest kind is used mostly to exclude injurious insects that may be in the plankton.

Other Foods and their Preparation

When entomostracans can not be supplied in sufficient quantities and of proper sizes for the young and mature fishes, it is necessary to provide substitutes drawn from the animal and vegetable kingdoms. In some special cases, certain of the other

Collecting minute crustaceans as food for goldfish

foods appear to serve a most useful purpose and are more or less regularly employed, but as a general thing the substitutes are of decidedly inferior value.

Mosquito larvae are acceptable food for older fish and, when cut up, for young fish as well, and they are often given. In the culture of the ranchu at Tokyo the brood fish for 10 days prior to egg-laying are freely fed with these insects. which may be collected in almost unlimited numbers in stagnant waters in all parts of the country.

Small annelid worms (*Tubifex, Limnodrilus,* and others) that live in the bottom of ditches and streams, and often occur in immense numbers, are frequently fed to the larger fish, and in Tokyo are particularly used for the brood ranchu during the week immediately preceding spawning.

Small fresh-water mollusks, especially gastropods of the genus *Viviparus*, are crushed and fed to the young during the summer following hatching when there is a scarcity of crustacean food. Another animal food that is sometimes employed is the silkworm, which is cultivated on a most extensive scale all over the southern part of the Japanese empire. The silkworms in the chrysalis stage are dried, pulverized, and mixed with some starchy material, and given to the fish in the first 2 or 3 months of their existence.

As food for very young fish the hard-boiled yolks of hen's eggs are rather commonly employed; and some breeders appear to prefer this to any other substance for the newly hatched fry of certain varieties, as, for instance, the ranchu. The pulverized yolk is mixed with a small quantity of water, strained through fine gauze, and distributed over the ponds by means of a watering pot.

Various kinds of cereal foods are used, either alone or in combination with the animal foods mentioned; among these are boiled cracked wheat and a mush made of wheat flour or corn meal. The smaller algae, particularly the unicellular forms, are often eaten by goldfish, but not from choice and not when other food is available. The), grow luxuriantly in the mud ponds, give the water a distinctly greenish color, and are indispensable in the cultivation of water fleas. Many are necessarily eaten incidentally, but they are not an efficient food and when taken to the exclusion of other things fail to promote a healthy growth.

Care of the Young Fish

For two or three days after hatching, goldfish remain very quiet on the bottom of the pond. They take no food through the mouth and require none, as they obtain all needed nourishment by the absorption of the yolksac. With the disappearance of the yolksac the fry begin to swim along the edges of the pond and to seek food. The earliest swimming efforts are feeble and clumsy, consisting of short spurts without any attempt at continuous movement. Incubation being completed, the materials that served as spawning beds are removed and the tiny fish are either transferred immediately to a clean pond or are retained for a time in the same pond, the practice varying somewhat with the locality and the variety.

Cardinal principles in caring for the young fish so that their growth may be favored and their colors developed at the proper time are to give them ample food adapted to their needs, to keep them warm and expose them to the sun's rays, and to renew the water in which they are held without subjecting them to sudden changes of temperature. Considerable experience is required in order to make the young take on their brightest colors at the earliest date. A novice may be surprised and chagrined to find that a given lot of fish will not exhibit any red color, while a part of the same lot in the hands of an experienced breeder will have completely changed.

At Koriyama, after the removal of the willow roots the ponds are drawn down and the fry are transferred to cement rearing ponds, in which the water has just been renewed. During the next five days there is no change of water; then fresh water is supplied, and thereafter there is a renewal about once a week as long as the fry remain in the cement ponds, which is usually for a period of 30 days after hatching. The smallest "mijinko" that can possibly be obtained by sifting are given to the fry during

Feeding entomostraca to young goldfish in concrete ponds

the first 10 days or 2 weeks of their swimming life; then larger ones, adapted for the increasing size of the fish, are provided. If for any reason the living natural food can not be given, the cooked yolk of hen's eggs may be furnished as a substitute.

Thirty days after issuing from the egg, the fry, having then reached the size of a rice grain, are transferred to a mud pond teeming with living food and containing no other goldfish and no destructive animals. The abundance of these crustaceans sometimes is extraordinary; the author has frequently observed rearing ponds in which the water was actually thick with copepods and other forms, so that the fry had only to open their mouths to obtain all necessary food. The pond is stocked with fry at the rate of 40 to 50 for the oranda and wakin or 20 to 25 for the ranchu to each square foot of surface. Should the supply of "mijinko" be inadequate for the rapidly growing fish, one or more of the various other foods mentioned must be provided. These are not scattered broadcast over and through the pond but are placed on shallow earthenware plates or trays, about 9 inches in diameter and provided with a rim, that are suspended by 3 cords from bamboo poles stuck in the bottom of the pond. The trays are arranged about the margin of the pond at depths corresponding with the movements of the fry. The very young fish can not withstand much pressure, and remain near the surface, so that the food is submerged only 1 or 2 inches; but as they grow they are able to go deeper and by winter take their food from trays 10 to 12 inches below the surface. The great advantage of this method of administering food is that the unconsumed portion may be withdrawn and is not left to decompose, pollute the water, and perhaps injure' the fish if eaten.

The foregoing description applies to the ordinary operations at a Koriyama goldfish farm. When the rearing is conducted in concrete ponds, as is sometimes the case with the ranchu, a daily change of water is necessary, and it is customary to use as the principal food the chopped larvae of mosquitoes (*Culex*). Given a concrete pond with a surface area of 100 square feet, there may be reared in it 100 ranchu under one year old, 30 under two years, 10 to 12 under three years. If, however, the number of fry is reduced one-fourth or one-third, the results are much better, and the fish may attain in one year a size that would otherwise require two years or more.

The fry of the oranda variety hatched in a concrete pond of 100 square feet need frequent change of water as long as they are held in their limited quarters, which is for about 40 days. They are then put in a mud pond where they are given ample room (8 to 10 fish to the square foot), which is increased by frequent sortings.

The great bulk of the stock is disposed of in the following spring. Fish intended to be reared to a very large size for the purpose of providing progeny of exceptional development of body and tail are fed freely on mollusks, silkworms, and starch; are given much room during their subsequent growing period; and are rigidly selected for their form and color. In their sixth year the space allowed them is 5 to 6 square feet per fish, and in the seventh year, after a further selection of breeders, the space may be 15 to 20 square feet per fish.

In the rearing of the ranchu after the Tokyo method, the first food administered to the fry is boiled yolk of egg, mixed with water and sprinkled over the pond. This food is given every morning for 7 days, the amount being determined by the appearance of the fry's abdomen, through the transparent wall of which the yellow food may be seen. Then for 15 days "mijinko" is provided, a sieve being used to exclude forms that are too large and noxious insects. Subsequently chopped annelid worms and mosquito larvae are given. During their active growth, the young are frequently sorted with reference to size or quality and put in other ponds, the number is gradually reduced by the elimination of the undesirable kinds, and greater space is thus provided for the others, so that by autumn the ratio of fish to pond area should be not more than 2 to each square foot. To protect these delicate creatures from the cold, a winter pond or hibernacle is provided in cases where the fish are not placed in a mud pond. The hibernacle is made of concrete, has a depth of 7 to 10 inches, is provided with a close-fitting lid, and has a peaked roof which on the north side extends to the edge of the pond. On warm days in winter the lid is lifted so that the sun's rays will enter the pond. If the water becomes foul, a gauze-covered basket is put in the concave area at one end of the pond and the water is withdrawn with a rubber syphon without any disturbance of the young, fresh water being then supplied.

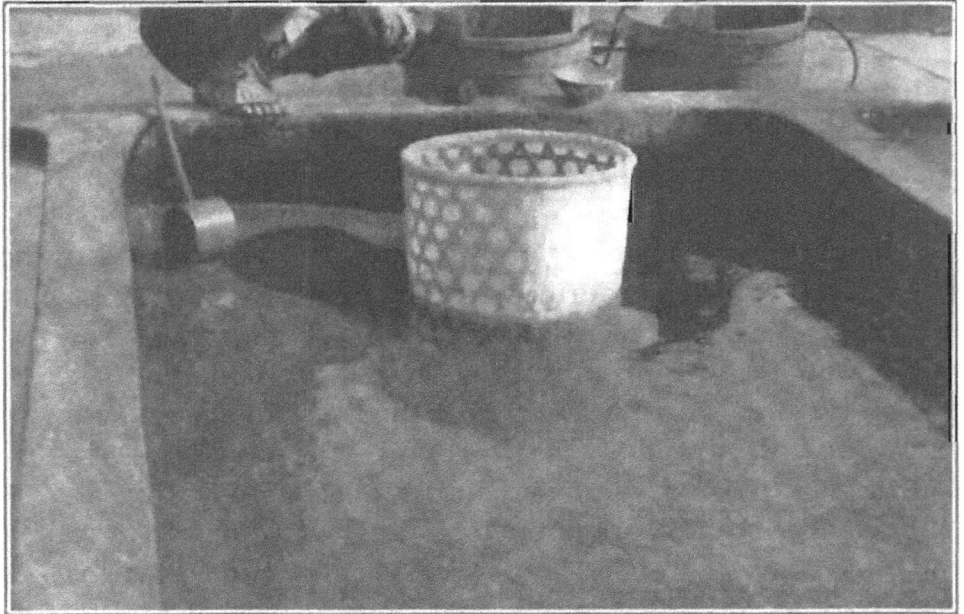

Drawing down a concrete rearing pond

The ryukin and wakin, being hardier than the ranchu, may be reared in mud ponds, to which the young are transferred 3 or 4 days after they begin to feed. For the successful cultivation of the ryukin, ample pond area is required for the fish and for the cultivation of the crustacean food. Supposing that 200,000 young are on hand, at least 4 ponds with an aggregate area of 7,000 square feet should be available for the alternate growing of "mijinko" and fish. With the exhaustion of the crustacean supply and the increased capacity of the young, mosquito larvae and other foods may be given on earthenware plates as previously mentioned. The demekin and the deme-ranchu are handled in the same way as the ryukin.

The young must at all times be protected from unfavorable

meteorological conditions. When the fish are in the shallow concrete ponds, frost and hail are much feared by the breeders, and hard and protracted rains also are dreaded. Protection against these agencies is secured by covering the ponds with a screen or canopy. When in the mud ponds, the young are injured by strong winds, especially in the early morning before sunrise, when the fish have the habit of coming to the surface and may be blown against the lee shore, banked up and killed. Winds therefore cause much trouble in spring and early summer, and necessitate all the attendants getting up early to provide against emergencies. If the wind is very strong bamboo sticks or poles are put in the ponds to break the force of the waves, and it may be necessary to net out the fish and deposit them on the protected side of the ponds. After an hour or less of sunlight the fish go down and the danger is over for the day.

Various enemies of the goldfish must be guarded against. Among these are birds, other fish. water snakes, turtles, frogs, and insects, the last being probably the most serious. A number of kinds of aquatic insects fly into the concrete and mud ponds from outside waters, and do much damage in a short time, pouncing on the fry and devouring them. It is therefore necessary to look over the ponds with these enemies in mind at least twice daily. Other diseases and fatalities to which the fish are subject will be noted later.

Sorting and Selecting the Fish

The goldfish varieties are very unstable, and exhibit a strong tendency to revert to the more primitive type and to show peculiarities of form and color that the culturists strive to keep in abeyance. This is particularly the case with the more recently developed varieties, and it is not strange that many of their

Transferring goldfish fry from one pond to another

progeny are not like their parents, but resemble the immediate or remote ancestors of the parents.

The pond being drawn down the young fish, concentrated in the circular depression, are dipped out with a bowl and deposited in tubs

From the earliest practicable date after hatching up to the time when full maturity is reached, the goldfish are subjected to repeated examination with a view to the elimination of the undesirable and unfit, so that the stock may be improved and the expense of caring for fish of little value obviated. The earlier selections are made with reference to form, while later both form and color are considered.

The selecting and sorting of the young in concrete ponds

are done by drawing down the ponds, dipping the fish into tubs by means of small bowls, then transferring them to white earthenware plates or trays, and finally taking them up a few at a time in small white-coated bowls and examining them with regard to the particular points under consideration. The fish being reared in mud ponds are caught with dip nets or with bamboo baskets lined with gauze, and are sometimes sorted on the spot and transferred directly to other ponds; to facilitate this a white earthenware platter is inserted in the net or basket as a background.

In the case of the ranchu, the Tokyo breeders and others whose methods are similar make many selections during the early stages. The first overhauling comes about 3 weeks after hatching when the young are so small that a sorting based on general characters is impossible, and only the size of the caudal fin is considered, the fish with the best development of this organ being picked out and put in a separate pond. At the second selection made 10 days later special attention is given to the symmetry of the body. After another interval of 10 days there is a sorting according to size, and 10 days later there is a final selection with particular reference to the shape of the caudal fin. With each examination the more desirable fish are given larger pond space; and after the third sorting the young, having reached a length of 1.2 to 1.25 inches, have a market value and those are sold that do not possess the desired qualities.

The practice of selling off stock of little value after the third selection is based in part on the circumstance that the fish destined to be white, and therefore most undesirable when grown, can then be distinguished from those that will develop a red or variegated coloration. By discarding the unfit and unprofitable the remaining fish have more food and room, and their growth and vitality are thereby promoted.

The sorting of the oranda shishigashira begins when the fish have been out of the egg 25 days and are about a fourth of an inch long. In the first examination, from two-fifths to two-thirds of the fry are discarded and sold as soon as possible: and 15 days later the reserve stock is transferred to a freshly-prepared mud pond at the rate of 8 fish to the square foot. The second sorting comes about the end of July, when the coloration may be more accurately gauged; at this time white fish and those with badly formed caudal fin are eliminated. Two other selections are made with reference to the shape of the caudal, and by autumn the number of fish remaining may be only 4 or 5 to the square foot. After being carried through the winter in the mud ponds without further sorting, all are sold in spring with the exception of those retained for rearing as brood fish.

The fry of the ryukin variety that have been fed for two -weeks from earthenware trays in the mud ponds are enticed into scoop nets or gauze-lined baskets by means of food, and as they are sorted they are put into recently flooded ponds. Selection at this age is based on shape of caudal fin and size of fish. Three to four weeks later there is another sorting according to size, and at the same time the fish without color are discarded. By the end of summer the elimination of the inferior specimens has proceeded so far that the ponds are stocked at the rate of 8 to 10 fish per square foot, instead of 60 per square foot at the outset. Early in the next spring all the fish are disposed of except 4,000 to 5,000, and these are subjected to still further examination from month to month until, of the several hundred thousand with which a culturist may have started, there remain for breeding purposes only 1,000 fish of superior grade.

Selecting and sorting young goldfish for rearing; much of this tedious work is performed by young girls

Growth and Color Changes

The rate of growth of goldfish is, within certain limits, largely a matter of food. The size difference between fish on restricted and unrestricted diet may amount to a hundred per cent in a given period. This is most important from a business standpoint, as the profits of goldfish culture consist chiefly in selling the largest possible fish at the earliest possible date. Growth is influenced also by the available pond space provided. The same factors likewise determine when the color changes occur, the brilliancy of the colors, the extent of the development of those special characters for which the breeders strive, and the reproductive capacity of the fish.

When grown in cement ponds that are fully stocked, the ranchu attains an average length (over all) of 1.8 inches at the end of the first year, 2.4 inches in two years, 3.6 inches in three years, 4.8 inches in four years, and 6 inches in five years. In ponds that are understocked, the fish may become 3 inches long in one year, with a corresponding gain in subsequent years. For more than a year the head is entirely free from the peculiar papillated growth so characteristic of the variety, and it is not until after the second or third year that this feature attains its full development.

The average size of the oranda at different periods is approximately as follows: One year after hatching, 1.6 to 2 inches, including caudal fin; two years, 2.4 inches; three years, 3.6 to 4.8 inches; four years, 7 inches; five years, 12 inches; six years, 13.5 to 15 inches. The warty growth on the head begins to develop about the second year.

At the end of the calendar year in which hatched, the ryukin under favorable conditions reaches a length of 3 inches, including the extended caudal fin; at the end of the second year the average length is 5.5 inches, third year, 6 inches; fourth year, 6.5 inches, and fifth year, 7 inches.

The ryukin requires at least four years to attain full development of its most attractive feature—the caudal fin.

In all newly-hatched fish the body is uniformly covered with a black pigment, which is an accompaniment of infancy and gradually disappears in the course of normal growth. The black covering begins to fade in the first summer and ordinarily within a year gives place to the adult coloration. The fish whose color changes earliest are likely to become white or white and red, while those that retain their original dark pigment longest are likely to be uniformly red. Many ryukins become striped before the initial color entirely disappears and leaves the body white. Fish that become entirely white meet with no favor and are always weeded out at the first opportunity, but a predominance of white may make very acceptable fish if associated with red fins and bright red blotches on head.

Some interesting facts, bearing on the question of heredity, have been brought to the notice of the goldfish breeders. The observations and developments have a rather important economic aspect, but have not been sufficiently correlated to warrant any general conclusions at this time in view of the complex, often unknown, ancestry of particular brood fish.

Because of the persistency with which the Japanese goldfish culturists endeavor to maintain and improve their stock by careful selection of breeders, it is noteworthy that even the long-established varieties have a most heterogeneous progeny, and in some instances breed true to the extent of only about 60 per cent. In the case of the oranda, a •Koriyama culturist reports that in his experience there may be expected 40 per cent of sports, consisting of about 30 per cent of fish having the form of the wild species, 5 per cent wakins with deformed tails, and 5 per cent ranchus, together with a few ryukins.

It is noted by Professor Matsubara that when ryukins with a two-rudder caudal fin are mated the progeny have long tails, but when fish with a one-rudder tail are brought together the tail in the offspring is short.

Sometimes, through the unfortunate choice of brood fish that appear to be entirely satisfactory, a large percentage of a lot of fish may have unsplit tails, whereas the culturist had every reason to expect split tails; this is serious from a business standpoint, as the fish with unsplit tail bring very much less money than the others.

Professor Mitsukuri cites it as an interesting fact that when the ryukin and the ranchu are crossed to make the oranda shishi-gashira both dorsal and caudal fins remain long, while in the crossing of the oranda with the ranchu to produce the shukin the dorsal fin is lost and the caudal remains long.

In all broods of the varieties lacking the dorsal fin, a certain percentage of the young show vestiges of that fin—sometimes a miniature fin, sometimes a few rays, sometimes a number of rounded protuberances, sometimes a single spine, all suggesting the comparatively recent period at which the fin was lost.

Transportation of Goldfish

The commoner varieties of goldfish bear transportation well if proper precautions are taken, whether the vehicle be steamship, express train, wagon or man, and whether the distance be long or short. Whenever there is serious loss in sending the hardier varieties from place to place, the responsibility usually rests on the shipper or attendant. Some of the more highly cultivated varieties, however, with feeble swimming powers and delicate constitution can be transported only with the greatest difficulty, and even with extreme care all or a large part of a consignment usually succumbs to long-distance travel. Hence some of the most interesting varieties have not yet become known outside of their oriental environment.

Prior to shipment goldfish, whether young or old, should be transferred to tubs, tanks, or small cement ponds with pure water and kept there without food for several days. This is to ensure the throwing off of all ingested matter throughout the alimentary canal. If fish are shipped with food in them, they are likely to die either because they will pollute the limited amount of water in which they are preferably carried or because the food will undergo decomposition in the intestines as a result of lowered metabolism.

The Japanese have learned by experience to use only a limited quantity of water in moving goldfish. It is a very important and suggestive fact that during either long or short shipments goldfish require only a very small amount of water, and the best results are obtained with the minimum quantity necessary to keep the skin and gills constantly moist. The writer has seen more than a thousand year-old fish carried by a man in two wooden tubs suspended from a shoulder bar, and this too in summer and for half a mile under a broiling sun. Notwithstanding there was not enough water to cover the fish, they were delivered without any loss whatever. Under the same conditions there would have been large mortality had an attempt been made to provide enough water to isolate each fish. The explanation is simple: The shallow tubs permitted the absorption of much oxygen from the air, and the absorption was increased by the squirming movements of the fish induced by the lack of water, the result being a plentiful supply of oxygen available for respiration while their gills and bodies were thoroughly moist—two requisites for existence.

For long shipments by water or railroad, goldfish should be kept in shallow tanks, tubs, or trays; and there should be available an ample supply of water for use from time to time as that on the fish becomes contaminated. The temperature of the air and water should be as cool as consistent with safety to the fish.

6 Diseases and Fatalities of Eggs and Fish

The high degree of domestication to which the Japanese goldfish have been brought predisposes to various maladies and fatalities from which the wild fish are for the most part free. The diseases of goldfish are due to bacteria, to animal and vegetable parasites, to improper food and feeding, and to physical causes, and are such as affect fresh-water fishes generally. The nature of some diseases of goldfish in Japan is not fully understood because not as yet investigated in a scientific way, and the treatment is largely empirical. While this subject is very important, it need not now be noticed at great length; and it will suffice for present purposes to mention some of the more common maladies, and to give the experience and views of Japanese goldfish culturists.

Under ordinary conditions only about 30 per cent of the eggs laid result in fish that survive long enough to reach a marketable size. The losses are 40 per cent during incubation and 30 per cent in the fry and fingerling stages. Losses of eggs are caused by frost, hail, and other physical agencies as already noted, but are due chiefly to fungus (*Saprolegnia*). Sometimes, owing to peculiar local conditions, many eggs become attacked by this troublesome growth and occasionally every egg in a given pond may be affected and killed unless proper measures are taken. The fungus does not ordinarily begin to develop on healthy eggs, but first attacks unfertilized or dead eggs and spreads thence to adjoining sound eggs. If affected eggs are removed as soon as noticed, the spread of the parasite may be checked and no great damage done; but if the water of a pond is thickly charged with the fungus spores and there is indication of a general infection, the healthy eggs should be removed to another pond filled with pure water, and the contaminated pond should be treated with sodium chloride, copper sulphate, or other fungicides, then drawn down and thoroughly cleansed.

Fungus attacks likewise young and adult fish, especially those with lowered vitality or with an abraded surface on which the spores may lodge, and eventually destroys if unchecked. When the characteristic white fuzzy growth appears, the healthy fish should be transferred to another pond, the diseased fish should be segregated in a tank or basin, and the infected pond should be drawn down and the bottom and sides thoroughly disinfected and dried. If the disease has not progressed too far the fish may be cured by the local application of a solution of common salt or peroxide of hydrogen, or by immersion in a moderately strong salt solution for a few seconds at a time or until the fish shows signs of suffering. The local use of other common antiseptics and fungicides (boracic acid, salicylic acid, formalin, carbolic acid, permanganate of potassium, etc.) in proper strength will naturally be suggested. Fungous disease is responsible for a large percentage of the mortality among goldfish, and should therefore be carefully studied by breeders and fanciers.

Crustacean parasites are common on goldfish as on all other fresh-water fishes. They are usually known as fish lice, and are found externally on scales and fins, and internally in gills and mouth. Those most frequently met with are copepods, which irritate when on the skin and fins, and may occur in the gills in such numbers as to seriously interfere with respiration and ultimately to cause death. Such parasites can usually be removed with tweezers.

The most serious diseases of goldfish are those due to bacterial and protozoan infections, which as a rule affect large numbers of fish and may destroy every fish in a pond or even a large part of the stock of a breeding establishment. When visitations come, attention can more profitably be directed to the removal of the cause, usually to be found in the water supply, rather than to attempts to cure the individual fish. The treatment of such diseases, unless superficial and localized, is most unsatisfactory; and as a general thing it is better to sacrifice the fish at once so as to prevent further infection.

One of the leading goldfish culturists of Japan has given to the writer the following memoranda of the diseases met with by him in the course of many years' experience: (1) "Kama." This is a serious epidemic disease affecting fish about ten days after hatching. The abdominal wall is thinner than normal and becomes attached to the dorsal side of the abdominal cavity. As the fish can take no food, they soon die. When the disease appears in a pond it spreads quickly, may attack nearly every fish, and may extend to other ponds. As soon as discovered, the pond is drawn down, drained, and cleansed, and all the fish are sacrificed, as there is no known remedy. In the year immediately preceding the author's visit, nearly every fish resulting from the first and second spawnings was lost in this way, and only the third and subsequent broods escaped the malady. (2) "Naginata." This affects also the young in the hatching ponds, and while not so disastrous as the preceding is much feared.

It is characterized by a swollen abdomen, and is thought to be caused by improper food.

The fish affected sometimes recover. The treatment consists in draining the pond, supplying fresh water, and feeding the smallest crustaceans in abundance. (3) "Kuchigusari." Fish in the hatching ponds sometimes develop a destructive inflammation of the snout, which sloughs off; the caudal and other fins also are destroyed. The cause is not known, and no cure has been found. (4) "Memuki." The principal symptom is a bulging of the eyeballs in fish a year or more old. The disease occurs mostly in April, May, September and October, and may be due to improper food or to failure to guard against marked temperature changes. The fish die if left alone, but may be cured by careful attention to the water and food. (5) "Chirosobu" and "Kurosobu." These diseases are characterized by white and black spots, respectively, on the body, and may be of a fungous or protozoan nature. The skin loses its luster and looks like Japanese paper. From one spot the affections may spread widely. They are thought to be due to lowered vitality owing to deficiency of food, and may be cured by proper feeding. (6) "Chinchiri." This name is given to a disease marked by swellings over which the scales project prominently; the swellings are soft and ultimately discharge a yellow fluid. Fish three or four years obi are most frequently affected. The trouble is ascribed to a sporozoan, and is thought to be incurable. (7) "Pest." A malady called the pest by goldfish breeders sometimes carries off many year-old fish. Black spots appear on body and fins, and there is much wasting of the muscles of the back. The nature of the disease is not known, but it is probably either bacterial or protozoan, and is highly infectious. On one occasion a pond in which the pest existed overflowed into a pond containing healthy fish, with the result that the next afternoon those ponds were badly affected. (8)

Swollen air-bladder. This trouble, which is not common, occurs in older fish and particularly in those living in deep ponds. The air-bladder being abnormally distended, the fish lose control over their movements and equilibrium, and float at the surface with the tail or belly upward.

7 The Standards of Quality, and the Trade in Goldfish

Fashions, Criteria, and Exhibitions

Owing to the age of goldfish culture in Japan and the great amount of attention devoted to goldfish by the general public as well as by breeders and fanciers, certain standards have been established, new fashions have from time to time been started, and individual and community predilections have been formed.

Considering the three leading varieties, it may be noted that the ranchu is in particular favor at Tokyo, the oranda at Koriyama, and the ryukin at both places. The oranda is most extensively cultivated in the southern half of Japan, where it has to a great extent replaced the ryukin. The variety was introduced at Tokyo but did not meet with a cordial reception, and is overshadowed by both the ranchu and the ryukin. The last is now grown most extensively at Tokyo, although up to fifty years ago the chief center for its production was Koriyama.

Qualities in the wakin that are considered desirable are a

thick, wide-spread tail with three or four lobes and fine rays, and distinct color pattern. Points of excellence in the demekin are symmetrical and strongly marked protrusion of the eyeballs, long and widely spread caudal fin, and a mottling of three or four colors, with conspicuous vermilion areas or black spots.

The highest type of ryukin has, as its cardinal feature, a perfect caudal fin; it may have either three or four lobes, and is long, slender, fine-rayed, soft, and pendulous; the peduncle is thick. The second point in determining the quality of a ryukin is the shape of the body and head; the body should be short and only two-thirds as long as broad, and the head should be broad and with a rounded snout. When the body is as long as broad, the fish ranks as second best. The least desirable examples are those with long body and short tail. Both back and belly should be variegated, and the caudal fin should be red. Fashion requires that the colors be not discreet but well blended.

The first mark of quality in the oranda is the caudal fin, which must be symmetrical, long, and flowing; at Tokyo and other more northern places a four-lobed tail is preferred, but at Koriyama and in the south generally a three-lobed, four-lobed, or bag-shaped tail is acceptable to fanciers. The shape of the head is the next important quality; the anterior part of the head should be broad. and the protuberances, according to Professor Matsubara, "should be like a large well-proportioned flower of the tree peony, and should not he small." Fish are rarely perfect in both respects—a well-shaped head will be accompanied by a short tail, and vice versa.

At Tokyo and by the adherents of the Tokyo school generally, the varieties chiefly cultivated are the wakin, ryukin, ranchu, and demekin; but the leading variety, and the one to which most attention is given by all persons interested, is the ranchu. There is much friendly rivalry among breeders and fanciers as to who

can produce and possess the finest specimens, and each year
in autumn there is held in Tokyo a ranchu exhibition at which
the claims of rival owners are considered by jurors and awards
are made. Professor Matsubara gives the following account of
this exhibition:

'The exhibition lasts two days, on the first of which are examined the grown-up breeds and on the second the young in
the first year. Being developed in color and form, the former
naturally attest the extent of skill in the breeders and their value
can be known at a glance. The latter are those hatched only in the
preceding spring and as yet little developed in every respect; but
these, after all the cares lavished upon them by breeders, are to
appear again fully grown up for contest at a future show, and on
that account are full of interest and promise. The examination of
fish in an exhibition is made in a shallow tub containing a white
earthenware plate in the center. Two ranchu being placed in the
plate are examined by connoisseurs as to their shade, dapples,
and the form of the tail and body. Those perfect in every respect
are awarded the 'first best,' and a list of the exhibits made in
the order of their merits is given to the public. Every time the
classification is made amid a stormy debate by the examiners.
No positive criteria exist to guide one in the examination of the
ranchu. Nevertheless, those uniformly bright red are considered
the best, so far as coloration is concerned. Those perfect in form,
however uniformly white they may be, are counted tolerably
good. The variegated ones at e generally unpopular. A ranchu
having either a white body and bright red tins and mouth, or
a bright red color in both cheeks, is also admired. Every one
of the breeds exhibited has its own name, which is given in
the aforesaid list with that of the owner. Those who participate
in the show are mostly nobles, wealthy merchants, and others
in comfortable circumstances. On such occasions the very best

breed fetches a price of two or three hundred *yen* ($100 or $150), but not one in ten thousand commands such a high price. Not a few goldfish breeders with fish of their own culture now come from localities lying far beyond Hakone to take part in a Tokyo exhibition. The reason why the exhibition is held in autumn is that the goldfish puts on the most brilliant colors in that season."

Recently a similar exhibition has been started in Osaka, where somewhat different criteria prevail. Thus, while a variegated ranchu is not regarded with favor in Tokyo, a fish with fine red and white mottlings and a bright red head is very popular in Osaka.

A noteworthy goldfish exhibit at the city of Sakai, held in connection with the great National Exposition at Osaka, was visited by the author. The choice fish, displayed in a long series of very shallow concrete ponds, were numerous as to individuals and represented most of the varieties known at that time. Those most in evidence and with particularly' fine examples were the maruko, the oranda, and the ryukin. One pair of marukos four years old and weighing probably half a pound apiece was valued at $50; the fish were white, with a few blood red blotches. Other specimens of superior breed had a nearly uniform golden body and red head. A beautiful type of ryukin was white, with the center of each scale red. Some large orandas 4 years old, with a conspicuous rounded mass of pink warts on the top of the head, were of the tokin or capped form; others were rich, velvety black, with a golden yellow suffusion on the under parts. Among the demekins were some comparatively large fish with small black blotches irregularly covering the led and white body color.

Sales and Prices

The goldfish industry is so completely eclipsed by numerous other branches of the fisheries that the Japanese themselves do not attach to it a great deal of importance commercially, and exclude it from the very thorough fishery statistics that are collected and published under government auspices. Therefore, it is not possible to present any figures showing the general extent of the business. In the Koriyama district at the time of the author's visit the normal annual goldfish crop was estimated at 10,000,000 fish. It is likely that the yearly production and sale of goldfish in the whole of Japan exceeds 20,000,000 and may reach a considerably higher number, and the aggregate value of the output cannot be less than half a million dollars.

Goldfish in Japan are so cheap that the poorest peasants buy them, and so dear that none but the wealthy can afford them. Inferior examples of the commoner varieties may be bought for half a cent apiece, which is probably less than is ever paid in any other country for any animated vertebrate ornament; while the most perfect specimens of the more highly cultivated varieties command higher prices than are given anywhere for any other kinds of fishes.

The ranchu is the most valuable variety, the oranda shishi-gashira closely follows, and the ryukin comes next. The wakin is the cheapest, and the other varieties have a value depending on their perfection and the local demand. The value of goldfish increases so much with their age that it is very desirable for breeders to keep their best fish until full maturity is attained; and in the case of the ranchu especially the practice is to retain the crop if possible until the fifth or sixth year, when the length is about seven inches, because such fish bring the highest prices.

In order to convey some idea of the actual and relative values of the different varieties of goldfish, the following average wholesale prices per 1,000 fish are taken from the operations of a Tokyo breeder a few years ago: Demekin, 1 year old, $10; wakin, 3 years old, $22.50; ryukin, 3 years old, $100; oranda

The selling house at a koriyama goldfish establishment

shishigashira, 5 years old, $750; ranchu, 2 years old, $75; ranchu, 5 years old, $2,500.

large part of the output of many breeding establishments is bought by itinerant vendors, who visit the ponds daily and take away the fish in shallow wooden tubs arranged in nests and suspended from a shoulder bar. The vendors do a particularly lively trade on holidays and festivals, but they find a steady demand at all times as they wend their way along the crowded streets and through the parks. One street seller seen at a Tokyo goldfish farm carried away at one time 500 goldfish of different kinds and sizes.

Fish awaiting shipment or collected for sale are held in bamboo baskets and live-cars moored in the mud ponds, or are exposed to view in the cement ponds. The export trade centers at Yokohama, Kobe, and Nagasaki, and the variety figuring most prominently in that trade is the ryukin.

8 Japanese Goldfish in America

Historical and Other Notes

The direct importation of oriental goldfish was accomplished at a comparatively recent date. The earliest lot of fish to arrive appear to have been those brought over by Rear-Admiral Ammen about 1878. From this stock came fish presented to Prof. Spencer F. Baird, United States Fish Commissioner, which were extensively bred from at the Fish Lakes in Washington. Later the importation of these fish was taken up as a business enterprise on the Pacific coast, and attained large proportions. The fish were first brought to the eastern States in commercial numbers by Mr. William P. Seal, of Delair, New Jersey, who for some years prior to 1894 controlled the output and supplied several thousands annually. At the present time one firm in San Francisco and another in Seattle are regularly engaged in bringing Japanese (and Chinese) goldfish to this country.

The variety that has been most extensively imported is the

ryukin, or fringetail. A few fish of the oranda variety have with-stood transportation and reached the hands of breeders; and in Philadelphia and elsewhere this variety has been success-fully bred. Several other varieties have been imported in small numbers. Owing to their delicate nature, the introduction of some of the most attractive and highly developed forms has not yet been accomplished, and a great treat is thus in store for American fanciers. Transportation presents serious difficulties which may never be overcome in the case of some varieties; but with increasing facilities on shipboard, shortening of the journey, and greater experience in handling, it seems likely that all of the Japanese varieties will in time be brought to the hands of our breeders. The acquisition of the most delicate forms may be indirectly accomplished by introducing their progenitors and breeding therefrom; the results of such crossing would with proper precautions, as hereinbefore noted, ultimately be the establishment of the desired varieties.

However, it seems probable that the importation of the tender varieties that now succumb to long journeys may be effected by the transportation of their eggs. By the use of a cool chamber, the slight retardation of egg development may result in the intro-duction of the most delicate forms, and this at less expense and trouble than the transfer of the fish themselves now occasions.

A very inviting field for the exercise of American ingenuity and skill is the opportunity for the production of new goldfish varieties of superior quality and exceptional interest by judicious crossing, rearing, selection, and cultivation of the Japanese forms. As the result of the hybridization of Chinese and Japanese gold-fish, American culturists have obtained a number of attractive, stable varieties to which the name Japanese has unfortunately been attached; but none of these can compare with the possi-bilities that are suggested by the further crossing of some of the Japanese varieties among themselves or with sonic of the forms that deserve to be called American. With such a plastic material on which to work, our breeders are certainly destined to bring into existence some noteworthy varieties—possibly the most remarkable that have been produced.

The furthering of this most inviting line of work, and the extension of goldfish culture in general, may be greatly facilitated by the formation of goldfish societies or guilds in all cities and towns. Such organizations, which are common abroad. particularly in Germany, would add a -most entertaining feature to local life and give to professional and amateur goldfish breeders and fanciers an opportunity to meet, exchange experiences, and hold exhibitions and sales. The flourishing Aquarium Society of Philadelphia. with more than 100 active members, does most excellent service for the promotion of goldfish culture in America and affords much pleasure and instructive pastime for its members. At the meetings, which are held monthly except in summer, there are special programs, discussions, and exhibitions, with award of prizes for the best specimens of goldfish in the different classes. The society has adopted a set of standards for judging the quality of the various goldfish breeds; and newly formed societies would do well to hold this older organization and its rules as models.

Goldfish Culture for Profit

The great and rapidly increasing interest in goldfish in America opens a wide and profitable field for professional goldfish culture in this country. There is no State where goldfish may not be grown. and there is scarcely a city, town, or section where goldfish cultivation can not be made remunerative.

The demand for goldfish is far in excess of the supply, and it is a common experience for dealers to be unable to fill orders.

This has been almost chronically the case in Washington and several other cities in the east, and probably the same has often been true of various other places where an effort is made to keep goldfish on sale. Furthermore, there are many cities and towns, to say nothing of smaller communities, where it is difficult, if not impossible, to obtain goldfish at any time.

A number of years ago the United States Bureau of Fisheries distributed Japanese goldfish gratis to applicants, and hatched and sent out thousands each season. This practice has long since been discontinued but there is a steady call for these fish from all parts of the country. The fishery service of the general government produces only small numbers of goldfish, that are intended solely for public ponds and fountains and for exhibition purposes; and private applicants are now referred to established dealers and breeders, of which there is only a limited number.

The best results in raising goldfish are attained when the ponds are in the open air; but American culturists have been quite successful with indoor culture or with a combination of indoor and open air operations. For outside culture on a large scale, the general methods of the Japanese should be followed, with such modification or adaptation as local conditions may require.

Where facilities are not ample for large mud ponds and for ponds intended primarily for the growing of crustacean food for goldfish, resort may be had to small cement or brick pools in which it is expected that no food will lie produced naturally. Such outdoor ponds, being shallow and not adapted for use during freezing weather, must b~ abandoned in autumn, and the fish must be cared for indoor s, in tanks or hothouses. Small hothouses or greenhouses, heated with an oil or coal stove. provided with glass top and sides, and fitted with a- series of cement basins, serve excellently for goldfish culture. They may be used in both summer and winter, but are particularly useful in winter because the fish will feed and grow during that season and be ready for market earlier than if kept out of doors. The temperature of such houses need not be maintained at over 50°F.

A goldfish breeder who desires to combine the useful and the ornamental may make his place very attractive by having his ponds form part of a landscape garden. The ponds may be on different levels, connected by little waterfalls. separated by gravel walks and greensward, skirted by trees and flowers, and provided with pond lilies, lotus, and other water plants; and the larger ones may contain islets reached by rustic bridges. Such ornamental arrangement need not be at the expense of any ponds required for practical culture operations, but, on the contrary, may very advantageously supplement the latter by supplying large picturesque ponds useful for wintering purposes, for producing natural food, and for rearing special broods.

A goldfish rearing establishment may be made one of the most interesting places in any community, and its financial success may depend in no small degree on the pleasure and instruction it affords visitors who may thereby he prompted to become patrons.

Suggestions for Maintaining Goldfish Aquaria in the Home, School and Office

A properly managed aquarium stocked with Japanese goldfish and provided with various incidental objects is one of the most attractive and instructive additions to a home, school room, or office. Goldfish may be maintained at much less expense and trouble than other ornamental animals occasion, and they should be very generally installed in residences, offices, and shops; and as an aid to nature study there should be a goldfish aquarium in every school.

Among the usual aquarial vessels, globes are the most unsatisfactory and undesirable. They afford greatly distorted views of their contents, and their contracted neck is objectionable because of the reduced water surface through which oxygen may be absorbed. As a general thing goldfish kept in globes fare badly and often are subjected to prolonged torture because of their cramped, poorly aerated quarters. Cylindrical glass jars are acceptable for this purpose, the best sizes being 12 to 15 inches in diameter and the same height. The best form of aquarium, however, is rectangular, with soapstone or slate bottom, and with four glass sides or with ends of stone. The smallest size to be recommended is 15 inches long, 9 or 10 inches wide, and 9 inches high. If room is available, the most satisfactory size is 20, 24, or 30 inches long. The width should exceed the height, and for the largest size mentioned the width might be 12 to 14 inches and the height 10 to 12 inches. Cylindrical jars and rectangular aquaria of all convenient sizes may be obtained from various dealers in all parts of the country.

The aquarium may rest on a window ledge, on a pedestal, on a stout table, or. if large, on a specially constructed support. It should be placed where it will receive ample light, and the direct rays of the sun should enter the water for a part of each day, but should not be admitted in such amount as to raise the temperature of the water unduly.

The use of running water in a house aquarium is usually unnecessary and undesirable. By observing conditions closely, it is possible to so adjust the various elements that an aquarium will become "balanced," and will maintain itself indefinitely without change of water. This is important and may be attained by adapting the number of fish to the volume of water, by securing a proper oxygenizing of the water through surface absorption and plant action, by having plants take up the carbonic acid gas resulting from the vital activities of the fish, and by providing for the removal of waste products (excreta, unconsumed food, decaying vegetable matter, etc.) by the use of animal scavengers and by periodic cleansing on the part of the attendant. Under ordinary conditions, fresh water need be supplied only to replace that lost by evaporation, the source of the water being immaterial provided it is clear, of proper temperature. not lacking in oxygen or containing injurious gases, and not strongly mineral.

In stocking an aquarium it is most essential that the number of fish introduced be no greater than the available supply of oxygen will easily maintain. Small fish ate preferable to large ones, because they require less attention and more can be accommodated; and those 2 to 5 inches long are the most satisfactory. For fish 3 inches long or less, there should be two-thirds of a gallon or one gallon of water apiece, while for specimens 4 to 5 inches long at least two gallons of water apiece should be provided, in a balanced aquarium.

To make an aquarium balanced or self-sustaining, it is necessary to introduce plants that are adapted to such an environment. There are many kinds of non-flowing plants that combine utility with beauty, and in almost every community water plants suitable for introduction into aquaria may be found in ponds, lakes, and streams. Among the most desirable are milfoil (*Myriophyllum*), hornwort (*Ceratophyllum*), eel-grass (*Vallisneria*), fanwort (*Cabomba*), pond weed (*Potainogeton*), swamp loosestrife (*Ludwigia*), and water weed (*Anacharis*).

Two or three kinds of the foregoing are sufficient for one aquarium, at one time, and the species may be varied at frequent intervals if desirable. A good combination is a floating plant with a rooted one. In a large aquarium some of the flowering plants— like the water hyacinth (*Piaropus*) or the arrowhead (*Sagittaria*)— may be inserted in limited numbers. Various filamentous algae are likely to be introduced incidentally with the other plants, and unicellular algae are certain to occur and at times may multiply to such an extent as to make the water green and turbid, thus obscuring the fish. Algae frequently coat the glass sides of aquaria, and may necessitate the cleaning, particularly of the front, in order to permit a proper view of the interior. A superabundant growth of plants will require removal of the surplus or a diminution in the amount of light.

Bottoms of house aquaria may be covered to the depth of 1 1/2 to 2 inches with gravel or clean sand, and the roots or stems of the rooted plants should be buried in the sand or tied to the pebbles to keep them in a natural, upright position.

Certain animals act as scavengers, and hence serve a very useful pm pose in aquaria when introduced in limited numbers. The best of these are tadpoles, and certain kinds of gastropods (snails) that will not attack the larger plants. Some of the snails consume the minute alga, and-serve to keep down the growth of these on the glass. In larger aquaria, one or two small specimens of mussels (*Unio, Anodonta,* etc.) may be allowed to burrow in the sandy bottom, and will add to the interest in the aquarium.

A glass cover will be found very desirable for every aquarium. It should not fit tightly but should be kept a quarter or half an inch above the frame by cork, wooden, or rubber buffers. Covers reduce evaporation, exclude dust and other foreign matter, prevent the fish from leaping out, and protect them horn- cats, rats, and other enemies. Some aquarists employ covers made of wire gauze or wire netting.

The water in a balanced aquarium quickly takes the temperature of the surrounding air, and hence its temperature varies greatly with the season and also from day to day. The goldfish may be subjected to a wide range of temperature without injury. Like other cold-blooded animals, so called, the goldfish assumes the temperature of the water in which it exists, and is able to adapt itself to 100°F. on one hand and 33°F. on the other, the essential conditions being that a change of temperature shall be gradual and that oxygen shall be present in the water in sufficient quantity. Moderately cool water, say of 50° to 70°F., is to be preferred to other temperatures. If the water is quite cold the fish are sluggish and less attractive, and if very warm there is danger from putrefactive conditions in the water and within the fish's intestines. Cold water is much the safer, as it has a much greater power to absorb and retain oxygen.

The losses which the amateur fancier necessarily meets with in the beginning are less likely to be due to neglect than to over attention. This is particularly true of feeding. Over-feeding and the use of improper foods are responsible for most of the losses in aquaria; where one fish dies of starvation, one hundred succumb to an overloaded digestive tract.

Inasmuch as the limited water supply of a home aquarium is quickly depleted of whatever natural food may be therein, from the outset it becomes necessary to provide food in quantity suited to the number and size of the fish on hand. In a properly appointed aquarium a certain amount of food will constantly be produced, but this is entirely insufficient; and dependence must always be placed on food from without.

There are various kinds of ready prepared foods suitable for goldfish in aquaria. Among these probably the best is the wafer made of rice flour that may be obtained from all dealers in ornamental fishes. A certain amount of animal food is essential, and the regular use of this will conduce to the growth and health of the fish. Readily obtainabl objects of this character are earthworms, that should he chopped into small pieces; mosquito larvae, that might easily he bred for this purpose; entomostraca, collected from an adjacent pond or ditch; ant larvae; and bits of oyster, mussel, and raw meat.

One of the pleasantest duties of the goldfish fancier is to feed the fish, and for the reason that it is interesting it is likely to be overdone, especially if all the members of a household undertake to give all kinds of food at all hours. Under the conditions in a small aquarium. goldfish do not require and should not receive food more than twice a day, and in some cases it may be better to give it only once a day. No definite rules can be given governing the quantity of food to be administered; this depends on special conditions, and must be based on experience and observation. It may be stated, however, that the amount of food should always be limited, and never in excess of the immediate needs and the actual consumption. If an aquarium is in such a place as to be affected by seasonal temperature changes, the fish should be given food less frequently and in smaller quantity during the colder months.

Food may be scattered over the surface and allowed to fall to the bottom, or it may be placed in a little tray or dish suspended a few inches below the surface. The advantage of the former method is that the fish have to seek their food and take it more slowly; the disadvantage is that the unconsumed food is not easily removable and the amount can not readily be estimated. The advantage of the latter method is the fish may be more closely watched when feeding, the proper amount of food can be more accurately determined, and the unconsumed food is not left to pollute the water -and can be easily removed.

A nicely balanced aquarium requires but little attention. So long as the water remains pure, no change is necessary; some of the most successful aquaria go for several years without a complete renewal of the water; and all that is required is to replace the water lost by evaporation. When water becomes foul through neglect or ignorance, the fish should be removed to another vessel, the plants thoroughly washed in running water, and the aquarium emptied, cleansed, and filled with pure water. When oxygenation is imperfect and the fish are suffering (as shown by their restlessness, evident discomfort, labored respiration, and efforts to get air at the surface), the water may be aerated by dipping it up and pouring it back from a height of a foot or more, flesh water may be introduced, and a more adequate plant growth should be provided. It is necessary to cleanse the sides of the aquarium of algal and other growths from time to time, in order to afford a good view of the fish. Accumulations of animal and vegetable waste matter on the bottom must be removed frequently—preferably each clay; this is done most conveniently by means of a glass tube about half an inch in diameter into which particles are drawn by first closing one end with the finger, or a flexible rubber tube used as a siphon. The siphon will be useful whenever it is desired to draw down the water.

In the limited quarters of the ordinary house aquarium, goldfish grow but little and may remain practically the same size for years. Spawning rarely occurs, and the propagation of goldfish in such a vessel is not to be expected. In the largest practicable home or school aquarium, if one or two pairs of medium size fish are kept and all the conditions are favorable, spawning may take place. As soon as the eggs are observed, they should be put into another aquarium or separated from the fish by a glass or wire mesh screen. The eggs and young are to be cared for as elsewhere described.

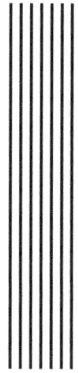

Appendix: Goldfish and Their Culture in Japan

By Shinnosuke Matsubara,
Director of the Imperial Fisheries Institute, Tokyo

Paper presented before the Fourth International Fishery Congress held at Washington, U.S.A., September 22 to 26, 1908

Japanese Varieties of Goldfish

That the goldfish of Japan was originally introduced from China there is no doubt, but a long domestication of the fish has resulted in producing Several Japanese varieties. The four known from remote times are as follows:

"Wakin" (Japanese goldfish).—This variety bears a close resemblance to the original *Carassius auratus* from which it came. Its body is slender and long, and the black pigment gives it when very young a color like that of steel, which gradually changes into vermilion red, often variegated with white. The caudal fin

is either simply forked or split into three or four lobes. (See page 11.)

"Ryukin" (Loochoo goldfish).—This variety has a short rounded body with a protuberant or swelled-out abdomen. The caudal fin, pendulous when at rest and flowing when in motion, is as long as the body or sometimes still longer, all the other fins being also long. (See page 15.)

"Ranchu," otherwise called *"Maruko"* (round fish).—The body of this variety is short and rounded, its tail and broad head being also short. It has no dorsal fin. The head, which is free from any abnormal features when the fish is quite young, in two or three years develops all over it a number of protuberances, like the achenia of a strawberry. In this state it is called "shishigashira (lion-headed) ranchu." Owing to the fact that this variety has a globular body, a short protuberant abdomen, and a short caudal fin, it can hardly swim, and is usually seen in an erect position with the head downward, which may be accounted for by the absence of the dorsal fin. (See page 19.)

"Oranda shishigashira" (rare lion-headed).—This resembles the ranchu in its bodily form with strawberry-like protuberances on the head. The body is big and longer than that of the ryukin and furnished with the dorsal fin; the caudal fin is long. (See page 24.)

Among the older varieties are also the following:

"Demekin" (goldfish with protruding eyes) and *"deme ranchu"* (ranchu with protruding eyes). The former was first introduced from China toward the close of the Japan-China war (1894–95). It has protruding eyeballs, and the body and the caudal fin are short. It is not usually bright colored, being black all over the body or yellowish red variegated with black spots or irregular patches. The deme ranchu, first brought from China six or seven years ago, has a globular body and, like the ranchu, no dorsal fin. The eyes not only protrude. but also are turned upward 90 degrees. (See pages 27 and 30.)

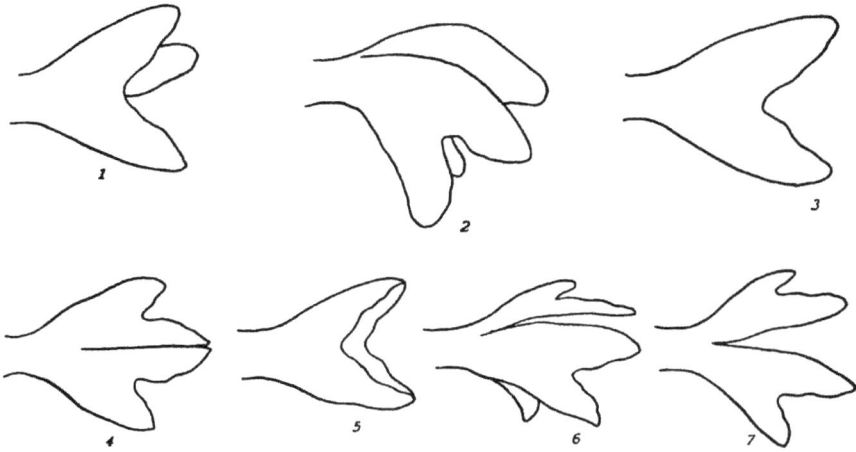

Typical forms of goldfish tails. 1: Arrow tail, lateral view. 2: Tassel tail, lateral view. 3: Forked tail, lateral view. 4: Three-lobed tail, dorsal view. 5: Bag tail, lateral view. 6: Rudder tail, lateral view. 7. Four-lobed tail, dorsal view

The foregoing are varieties for some time cultivated in Japan or recently introduced from China. New varieties, however, different from these, have lately come into existence, and I shall now describe how and in what order these have been produced.

"Watonai."—When the Fisheries Exhibition was held in Tokyo in 1883, I saw there exhibited a highly interesting variety of goldfish, which was a wakin with a flowing caudal fin. Being struck with its extreme peculiarities, and inquiring how this variety came to be produced, I found that the exhibitor, who lived at Imaidani, Akasaka, Tokyo, had only one pond for breeding goldfish, and that there the said variety was produced. As in that pond all the different varieties just stated above were reared together, the strange new variety appeared to be a cross between the ryukin and the wakin; but I did not make at that time further investigation into the cause. That breed was called "watonai,"

which means "a variety hitherto found neither in Japan nor in China." (See page 31.)

As to the order in which the different varieties of goldfish have been produced in Japan, it is probable that the wakin stands first in priority, but in what order the others have come can not be definitely stated. The oranda shishigashira, however, is generally known to have been produced by crossing the ryukin with the ranchu at Koriyama, Yamato, in 1840. All the new varieties of recent times except the watonai, which comes first in order, were produced by Mr. Akiyama Kichigoro, a goldfish breeder in Tokyo. The latest produced varieties are the following three:

"Shukin."—In 1897 a new variety, having its own peculiarities, was produced by the following means: Oranda shishigashira, ten in number and 3 years old, were crossed with ranchu equal in number and also of the same age. The cross thus made produced about 300 young fish, of which some were like the oranda shishigashira in form. There were 20 entirely without the dorsal fin; but in a large majority of the young there were some traces of the dorsal fin, one to three spines or something like knobs being discernible in its place. Those without any dorsal fin were then selected for breeding purposes, and in their offspring were still found some with traces of dorsal fin such as the spines and knobs mentioned above. This variety without the dorsal fin has been kept breeding up to the present; and being 'requested by the breeder to give it an appropriate name, I called it "shukin" (literally "autumn brocade"). The adjective "shu" takes its rise in the breeder's name, Akiyama (Chinese, Shu-zan, e.g., "autumn hill"), and the "kin" signifies "brocade," the epithet being given on account of the beautiful bright color. (See page 32.)

"Shubunkin."—The year 1900 first saw this new variety. The goldfish hitherto known in Japan having no such dapples as are found in the Chinese demekin, which is dappled in three or

four colors, the breeder had wished to have a variety produced having the same dapples as the Chinese. Thus the males and females, each ten in number, of the demekin having the most attractive dapples were crossed with an equal number of the males and females of the wakin. The breeding fish selected were the demekin having black dapples on a vermilion red or purple ground and those speckled with red, white, black, and blue. Some of the offspring resembled the original *Carassius auratus in* form and had dapples like the demekin, while in some others the scales were not so conspicuous as is the case with the demekin. On the whole, however, those resembling the wakin were most numerous, while those similar to the demekin in form were very few; and the number of the cross the breeder had in view was only 100 out of a total of 500. This new variety, to which I gave the name of "shubunkin" ("vermilion red dappled with different hues"), has since been kept breeding. Among its descendants some were of a vermilion color, dappled with black, some of a purple color all over the body without any dapples, and the others speckled with three or four colors. The entirely purple color above referred to was quite an unknown thing in the parent fish. (See page 34.)

"*Kinranshi.*"—With a view to obtaining a new variety without the dorsal fin, the ryukin was crossed with the ranchu. The number of parent fish, as well as the numerical ratio of the males to the females, was the same as in the preceding case. The offspring thus produced were of a bright attractive color. Some were provided with the dorsal fin or spines, others partially provided with the former, while in some others there were protuberances in place of the dorsal fin. Those entirely destitute of any dorsal fin numbered only one-third of the whole. This last-mentioned variety, which I called "kinranshi" (brocade-figures), has been kept breeding up to date. (See page 35.)

Of these three new varieties, the shukin is the most popular, but the shubunkin the most profitable one, only a few of the latter being eliminated on account of the caudal fin being crooked or irregular.

The goldfish is cultivated almost all over the Empire, but most extensively in Tokyo and Koriyama, Yamato, Nara Prefecture. Broadly speaking, the most popular varieties are those cultivated in these two localities, and the methods of rearing goldfish most in vogue are also those followed there. The northeastern half of the Empire follows Tokyo in things relative to the goldfish breeding, while Koriyama leads the southwestern half.

Goldfish Culture in Tokyo

The varieties of goldfish cultivated in Tokyo are the ranchu, the ryukin, the demekin, and the wakin. Attempt was once made in Tokyo to cultivate the oranda shishigashira by introducing the fish from Koriyama. This variety was not regarded with much favor, however, and has come to be scarcely ever cultivated in the capital. The most popular breed is the ranchu, which is highly prized by goldfish lovers and engrosses a great deal of the attention of breeders, who take pride in producing the best of this variety.

Ranchu

This fish is chiefly cultivated in concrete ponds. When the culture is carried on upon a small scale, the usual number of parent fish is five, two of which are females and the others males, but when it is carried on upon a large scale, the number of parent fish is over fifty, one half of these being male and the other half female. The eggs are laid during the period between the beginning of April and the middle of May, and the utmost attention should be paid to the fish during the three months of September, October, and November of the preceding year, when the food should be given them in sufficient quantity, without their being overfed. The males of the parent fish are separated from the spawners some time before the latter deposit their eggs. When the spawning season approaches, the water of the pond is not renewed, but the fish are kept amply fed with the larvae of mosquitoes (*Culex pipiens*) or earthworms (*Limnodrilus* or *Tubifex*) for about ten days, at the end of which time the eggs are laid when the temperature of water in the pond rises or when it rains. As, by experience, this can be known beforehand, the water of the spawning pond is changed the day before and the parent fish are removed thither. The spawning bed is then made with "kingyomo" (*Myriophyllum verticellatum*), on which the eggs are deposited on the following morning.

When the number of parent fish is 5, the proper size of the pond is 4 by 3 shaku (1.2 meters by 91.0 centimeters) with a depth of 5 sun (15 centimeters). With so parent fish, the area of the pond should be 7 by 6 shaku (2.1 by 1.80 meters), the depth remaining the same as before. In a pond of the latter size, a circular concavity with a diameter of 8 sun (24 centimeters) to 1 shaku (30 centimeters) and with a depth of 1 sun 5 bu (4.5 centimeters) from the bottom of the pond, is made around the outlet for the water. This concavity is intended to prevent the fish from being driven, when the water is drained off, against the walls of the pond. When the spawning is over, either the parent fish or the eggs are removed to another pond to prevent the eggs from being devoured.

For a few days after the eggs are hatched, the fry do not move about, but stick to their bed or the bottom of the pond. During this time no food need be given, as every individual of the fry is provided with a yolk sac in its abdomen; if forcibly fed, their health would be impaired. In three days or so, they begin to swim about in the pond, and are then fed once every day in the morning with the yolk of hen's eggs boiled. This food is administered by straining the yolk, mixed with water in the proportion of 2 eggs to 5 sho (9.1 cubic centimeters) of water, through cotton or silk gauze, this mixture being then completely stirred up until it is yellow, when it is put into a watering pot and poured all over the pond as food. The belly of the fry becomes yellow by taking this food, and the shade of the abdominal yellowish color shows whether or not the fry have got sufficient food. After being thus fed for seven days, they are given the small crustacea, "mijinko" (*Daphnia, Moina* or *Cyclops*), which have been cultivated and kept in a separate pond, the crustacea being caught with a gauze bag and then sifted, in order that any injurious insect, etc., may not remain mixed in them. After fifteen days thus nourished with mijinko, the tender fish are fed with the larvae of mosquitoes or earthworms cut in small pieces. About twenty days after the fish are hatched, the first selection is made, for which purpose the fish are put into a white earthenware plate. As they are still very small at this stage of development, it is difficult to select those which are satisfactory in every respect, and only those furnished with the better caudal fins, such as they are, are picked out. Ten days later the second selection takes place, when those having any irregularities on the back are thrown into a mud pond and put aside. Again, ten days later, the third selection is made and the fish are grouped together according to their size. And yet, about ten days after this, the fourth and last selection comes, when particular attention is paid to the form of the caudal fin. The fish are for the first time put up for sale when the third selection is over.

After the first selection the young are put into two concrete ponds, each of which is of the same size as the former one, and in the second selection they are distributed into three ponds of this same size. The area of these ponds would not be large enough for the original number of fish, but the number has been reduced as the inferior breeds were eliminated. To every twenty of the fish hatched in the beginning of May is given a space of one tsubo (3.3058 square meters), and when the cold weather comes they are removed either to a mud pond or to a hibernacle (i.e., wintering pond). The latter, which is made entirely of concrete and has a depth of 6 to 8 sun (18.2 to 24.2 centimeters), is provided with a lid, and the whole is again covered with an inclined roof opening toward the south, with one end coming down to the north edge of the pond. When the weather is warm, the lid is partially lifted up so as to admit the sun's rays into the pond from the south. The water sometimes becomes foul through the putrefaction of fish food and from other causes, in which case a basket having its lower part covered with gauze is immersed into the concavity of the pond and by means of a rubber siphon put into the basket the foul water is drawn out, to be replaced by fresh water. The fish in the second and third years should have a space of one tsubo at least for every twenty of them.

In the first year the fish attain in eight or nine months the size of 2 sun 5 bu (7.5 centimeters) from the snout to the extremity of the tail; in the second year they attain in the same length of time the size of 4 sun (14 centimeters); in the third year, 5 sun (15 centimeters); in the fourth year, 5 sun 5 bu (16.5 centimeters), and in the fifth year, 6 sun (18.2 centimeters). The fish that are kept four or five years are the only ones fetching a very high price in the market.

Ryukin

Being naturally very strong and healthy, the ryukin is much in demand abroad and is exported in large numbers. At home, also, this variety is a great favorite, and as such is kept in the ponds. Those having the longest caudal fin and called "ohiki," viz., "tail trailers," require at least four years for their full development. The superior kind has a head broad in front but not angular, and a short body. For instance, to be perfect in form, the body should be about 1 sun (3 centimeters) in length when it is 1 sun 5 bu (4.5 centimeters) in breadth. The body as long as broad comes next. As to the form of the caudal fin, it should be long and thin, at the same time the rays being fine and slender and the peduncle of the tail thick. To have the best coloration, it is essential that both the belly and the back should be well dappled and the tail vermilion colored. In former times when goldfish were kept in a china basin to be looked at from above only, those having the finest dapples on the back were most highly appreciated. Nowadays, however, the fish are kept in a glass tank to be viewed from the sides, hence the necessity of having fine dapples on the sides and downwards. The greatest desideratum for the ryukin is that it should have nothing abnormal in the caudal fin.

When the fish having a two-rudder tail are selected for breeding purposes, the tail becomes long in the offspring, but when the fish having a one-rudder tail are used for the same purposes, the tail becomes short in the offspring. Those having a long body and a short tail are regarded with the least favor. The second requisite for the ryukin is that the form of the body should be satisfactory, the shape of the head coming last.

A mud pond is used not only for rearing the ryukin but also for its spawning. Equal numbers of males and females of this variety are used as breeding fish, and 800 fish 3 years old are placed in a pond having an area of forty tsubo (132.2 square meters) with a depth of 2 shaku 5 sun (76 centimeters). An ample supply of food is given them in the preceding year. In order that sufficiently developed offspring may be put up for sale at the earliest possible date, the breeders vie with one another in causing the parent fish to deposit their eggs with little delay. Many of the offspring become striped like a tiger before their original steel color fades into white, and it is at this intermediate stage of their coloring that those inferior breeds destined to become white can be sold with the greatest profit. A full supply of fertilizer is required in the mud pond in which the breeding fish are kept, as it produces more plankton for them. About the end of March, when its temperature rises to 15° C., the water of the pond is renewed. In this season of the year, when the weather is very changeable, the fish require the utmost care. It is good for them if the temperature of water in the pond is higher than 15° C., but a lower temperature has a bad effect. A rising temperature accelerates the hatching of eggs, but a falling one retards it from two to three days, thereby producing a diversity of size in the offspring, an effect that should be avoided. When the eggs are deposited thereon, the bundles already referred to are removed to one or two ponds. In the case of its size being 4 tsubo (13.2 square meters), one pond may suffice to receive the eggs, but in the case of its being only 2 tsubo (6.6 square meters) the eggs are distributed into two ponds. The more spacious the pond is, the better it is for breeding purposes.

The best food for the young is natural food, viz., plankton, which is on that account cultivated beforehand. About as much as one koku (5.2 bushels) of rice bran, soy lees or the excrement of man or horse is put into the pond, if it is so tsubo (165.3 square meters) in size and over Io years old. But if the pond is only about 2 years old, a double quantity of the fertilizer is used. In each case, the pond is drained previous to putting in the fertilizer, after which its bottom is exposed to the sun's rays for about a week, and then the pond is filled with water. Another pond of 40 to so tsubo (132.2 or 165.3 square meters) in size is provided for receiving the fry just hatched. This pond already contains the plankton which was produced by feeding the pond with about 3 to 4 to (1.5 to 2.1 bushels) of the aforesaid fertilizer. Besides, if 1 go (0.005 bushel) of mijinko is put into the pond three days prior to removing the fry there, they increase to an immense number in that short interval.

When the young fish begin to swim about in the pond, the bundles on which the eggs were deposited are removed, and they are fed from the following morning with boiled eggs (5 in number), prepared as in the case of the ranchu. Three or four days later, when it is warm, the water in the pond is drained off by means of a rubber siphon, as stated before, and the young are put into a mud pond and fed with mijinko. Ten or fifteen days after, the fish, having outgrown the size of the pond, have to be removed to another pond. When a supply of mijinko falls short in that pond, the fish are again put into a new pond where plenty of such food can be had.

In order that the ryukin may be reared with any success, at least four ponds are required, two of which are intended for keeping the young and the other two for cultivating mijinko. In the middle of May the fish are grown so large that the larva of mosquitoes can be given them as food on earthenware plates slung by three strings from a bamboo pole. Given 800 parent fish, 200,000 offspring are produced, which is a proper number for the capacity (200 tsubo) of the four ponds just mentioned.

No selection of the fish is yet made at this stage of their development (the middle of May). After feeding them for about fifteen days from earthenware plates, the selection is first made in the following way: Food being put into a basket or a scoop net, the young are thereby enticed and caught, and are distributed into two ponds. The first pond is at the same time completely cleaned from everything pernicious and replenished with water.

The selection is sometimes made while catching the fish with a basket or a scoop net from the new pond, when an earthenware plate is put for this purpose into the basket (or the scoop net), the inside of which is covered with gauze. The caudal fin is the criterion by which selection or rejection is made. If there is any marked diversity in the size of the fish, the large ones are separated from the small. Those selected are put into the pond lately cleaned, when their number is roughly estimated by measuring them with a teacup or a lacquered bowl. Twenty or thirty days later another selection takes place, in which the fish are chiefly classified according to size. As at this stage those destined to be white can be distinguished from those to be dappled, the former are caught and sold off. Between this time and the following August all the fish are cleared off by selling, except 300 per tsubo of the ponds, it being possible to find buyers for the fish grown to one sun (3 centimeters) or over.

Toward the end of March the next year, 4,000 of the fish deemed best are retained and the rest sold off. From that time forward selection is made once a month during the following April, May, and June, and every time, 1,000 fish are eliminated, commencing with the most inferior, until 1,000 are left at last for breeding purposes. Making allowance for loss from various causes and deducting 2 per cent from the above number, 800 fish may remain.

Wakin, Demekin, and Deme Ranchu

The wakin is never so much prized as the ryukin, but on account of its large size, besides being healthy and strong, it is kept in a garden pond. In this variety, the tail should be thick and widely spread, the rays invisible and the dapples not merged into one another, while the contrary is the case with the ryukin in this last respect. The method of its culture is the same as that of the ryukin.

Of the demekin those dappled with three of four colors are highly prized. Either the vermilion dapples or the black patches should be pronounced. In some the black pigment gives the color of ink, while others have the color of steel, like that of the crucian carp. In the former the color is permanent, but in the latter it fades into yellowish red in a year or two, and such a color as this last is worse than a uniform red. The caudal fin should be long and widely spread. The eyes, right and left, should protrude symmetrically. Those which protrude but little are not regarded with favor. Goldfish of this variety swim about, not in groups, but singly, which is not the case with the goldfish long known in Japan.

The deme ranchu is colored all over the body either with yellowish red or black pigment, or yet dappled with black and red. This variety remains most of the time at the bottom of the pond resting on the belly, and scarcely ever swims. It does not live in groups; even less so than the demekin, and it very seldom spawns. In cultivating this variety, the breeders follow the same method as with the ryukin. Thus far as regards the culture of goldfish in Tokyo. I shall now describe the method of culture pursued at Koriyama.

Methods of Culture in Koriyama

There are but few trustworthy records giving information on goldfish breeding in remote times in Japan. Tradition has it that during the Hoyei era (1704–1710), a certain Sato Sanzaemon set up as a goldfish breeder at Koriyama and commenced to cultivate the fish in a mud pond. It is said that he pioneered the industry in that locality.

The principal varieties of goldfish cultivated at Koriyama are the oranda shishigashira, the wakin, and the ranchu. The ryukin was most extensively cultivated there until fifty years ago, but it has since been replaced by the oranda shishigashira and is not much reared at present. The demekin is not wholly uncultivated at Koriyama but no great attention is paid to it.

The dimensions of a pond at Koriyama do not vary with the different varieties of goldfish. It is usually oblong in form and measures Io ken by 3 ken, i.e., 30 tsubo (99.2 square meters). This is a size most convenient for the purpose of efficient fertilization. Formerly a depth of 1 shaku 5 sun (45 centimeters) was given to the pond, but it is now increased to 2 shaku (60.5 centimeters) in order to protect the fish from the sun's heat or atmospheric changes. The only drawback in this increased depth is that injurious gases are thereby generated from the bottom, especially when a large quantity of fertilizer is put into the pond.

Collection and Cultivation of Fish Food

The most noteworthy thing in connection with goldfish breeding here is the attention paid to rearing as well as collecting mijinko. These small crustacea are caught in a bag 20 to 25 shaku (6.0 to 7.5 meters) long, with a diameter of 2 shaku (60.5 centimeters) and made of "tenjiku kanakin" (a kind of calico) with fine meshes, varnished with the astringent obtained from unripe persimmons, or with the extract of oak-tree bark. For two weeks after hatching the fish are fed with the smallest of mijinko, which have been sifted. For another fortnight a larger kind of the crustacea is also given. When mijinko can not be obtained recourse is had to the yolk of boiled eggs, well pulverized. But the latter compares unfavorably as a substitute.

Usually 40 days, but when it is warm, 30 days, prior to removing the newly hatched fish to another pond, soy lees packed in a straw bag are immersed in the bottom of the second pond with a view to producing mijinko. To determine properly what quantity of fertilizer is to be placed in the pond is very difficult. In the case of a newly made pond, 50 kamme (187 kilograms) of soy lees mixed with 4 or 5 ka (59.8 kilograms) of human excrement in liquid state are poured over the bottom when it is not yet filled with water. When the bottom is sufficiently exposed to the sun, the pond is replenished with water, which in a few days presents a green color on account of the algae produced there. Even after the young fish are put into the aforesaid pond it must be fertilized every other day. When they are fed with mijinko obtained elsewhere, one sho (0.4 gallon) of the crustacea is given every day in the case of a pond of 30 tsubo (99.2 square meters).

This is what should be done with the fish when removed to a new pond after forty days from the time of their hatching. During these forty days they are kept in a concrete pond and are given mijinko in such quantity as may be deemed proper. Afterwards the number of mijinko in that pond is ascertained by immersing something like a white earthenware plate in a corner of it, and a supply of food to the fish is kept daily increasing. Feeding the pond with fertilizer is intended for no other purpose than producing mijinko, with which the fish are to be provided without interruption.

The longer the fish are fed with mijinko the better it is for them. As there is, however, a limit to the supply of mijinko, that wholesome food when the young are grown to 4 or 5 bu (1.2 to 1.5 centimeters) gives way to the Viviparus (a kind of mollusk found generally in the rice fields) pulverized, or the dried chrysalides of silkworms pounded and mixed with starch. As it is during the three months of June, July, and August that the fish increase in size with great rapidity, the most abundant supply of food is given in this season of the year. The extent to which the fish are to be fed, indeed, is judged by the color of water in the pond. When the water is green and turbid, it shows that the supply of food is plentiful; when it is green and transparent, the supply is insufficient. The same is the case with all other colors.

Oranda Shishigashira

A concrete pond is used for hatching the oranda shishigashira. It is 12 shaku (6.1 meters) long by 5 shaku (1.5 meters) broad with a depth of 7 to 8 sun (21.2 to 24.2 centimeters). The eggs having been deposited on bundles of willow-tree roots placed in a mud pond, about 50,000 or 60,000 are put into the above-mentioned concrete pond to be hatched. The water is removed without fail every four or five days, and is even changed every other day when the weather is warm. The first selection of the fish takes place after twenty-five days from the time of their hatching, when they are about 2 bu (0.6 centimeter) in length. About 20,000 of the then superior breeds are retained, the rest being eliminated, i.e., sold off. If the number of those first hatched suffices, it is most satisfactory, but if it does not, those hatched later are added. If sickness or other cause reduces this number, those hatched still later are used also. In that last case the fish are made to spawn even in midsummer. Forty days from the time of their hatching they are put into a mud pond in the proportion of 300 fish per tsubo. The next selection is not made until the fish put on new colors at the end of July or the beginning of August, when those which are uniformly white or unsatisfactory in the form of the tail are sold at a nominal price. That the best kind of the variety may be obtained, however, two more selections are to be made, the character most highly prized being the form of the caudal fin. In Tokyo and Nagoya the four-lobed tail is regarded with preference, while Koriyama is content with the three-lobed tail, the four-lobed tail, or the bag tail, if only it is symmetrical in form. The head should be broad in front. The protuberances which are in three years developed on the head should be like a large well-proportioned flower of the tree peony and should not be small. In some localities, such as Kii, Hiroshima, Awa, and Sanuki, there is a variety called simply "shishi" (lion), which is brass colored and has a short tail. Not a few of this variety are found to be uniformly colored. Generally speaking, those which are satisfactory in the form of the head have the tail abbreviated; seldom are the fish perfect in both respects.

There was formerly neither oranda shishigashira nor ranchu having variegated figures on the back. Such varieties, however, have been occasionally produced since the twenty-second year of Meiji (1889), and their descendants have been studiously used for breeding purposes. Both varieties are now extensively produced and highly admired.

The young oranda shishigashira are first put into a mud pond in the proportion of 300 fish per tsubo, but those white in color and abnormal in form are eliminated, and from this and other causes the numbers are by autumn reduced to one-half of the number originally put into the present pond. These are left alone till the following spring, when they are put up for sale. They attain by that time the size of 1 sun 5 or 6 bu (4 to 5 centimeters) in length. When it is desired to produce fish of a greater size, the number of young first to be removed to a new pond after hatching is reduced from 300 to 100, and these attain the size of about 2 sun (6 centimeters) in the same length of time as before. In the third year they grow to 3 or 4 sun (9 to 12 centimeters), in the fourth year, to 6 sun (18 centimeters), in the fifth, to 8 sun (30 centimeters), and in the sixth, to 1 shaku (33.3 centimeters). In fact, it is known that in the last-mentioned year they sometimes attain to the minimum size of 1 shaku 2 sun (36.5 centimeters). Furthermore, for the purpose of producing breeds of a great size, the fish having a good natural constitution and a well-formed head are selected from the six year old stock as parent fish, and breeds are put into a pond in the proportion of 6 or 7 per tsubo. Selection is again made in the following year and the rate is further lessened to 2 per tsubo. The fish are fed with 2 kamme (7.5 kilograms) of Viviparus and 500 or 600 momme (1,875 to 2,250 grams) of the chrysalides of silkworms pounded and mixed with starch (of wheat). All ponds are so made that they can be drained at any time to prevent the generation of poisonous gases. Usually twice a year, viz., in March or the beginning of April, when the young are about to be removed to another pond, and at the end of autumn when the fish are going to be put into a hibernacle, both these ponds, prior to receiving them, are drained and dredged and then exposed to the sun's rays for four or five days. Any place where the water gushes forth in the ponds should be exposed longer. It is usual for these steps to be taken twice a year as stated above, but the more this is done the better it is for the health of the fish; it would be best to do it even once a month. When the fish come up to the surface of the pond to breathe in warm weather before sunrise and go down afterwards into the water it shows that they are in good health.

Ranchu

The ranchu is chiefly cultivated in a concrete pond, though in small numbers. The number of ranchu to be put into a pond after hatching is one-half the number of oranda shishigashira, and the amount of fertilizer used is also half the amount of that used in the case of the latter variety. When put into a mud pond, the method of culture of the ranchu does not differ from that for the oranda shishigashira, but when reared in a concrete pond it is essential that the fish should be constantly supplied with fresh water. Hence the necessity of entirely renewing the water once every day. An amateur breeder would be likely to partially change the water to prevent a sudden change of temperature in the pond; but nothing is better than entirely replenishing the pond with fresh water of the same temperature as before. The fish are fed only with the larva of mosquitoes.

In a concrete pond of the aforesaid size are generally placed 100 fish under 1 year, 30 under 2 years, 10 under 3 years, 4 under 4 years, or again 4 under 5 years, but if these numbers were reduced better results would be obtained. It is good for the fish to be constantly supplied with food. Earthenware plates are not used for this purpose as in Tokyo, except for the time being after the fish are first put into a mud pond. There is one advantage in this method of feeding them without earthenware plates: they are naturally made to thrust their snouts in quest of food into the bottom of the pond, with the result that the bottom remains free from gases. If the number of fish under 1 year to be put into a rearing pond is reduced below the normal quantity above referred to (i.e., 100), they can be made to attain the size of 2 sun 5 bu (7.5 centimeters) at the end of the year, while keeping to that quantity they attain in length only 1 sun 5 bu (4.6 centimeters) in the same length of time, 2 sun (6 centimeters) at the end of the second year, 3 sun (9 centimeters) at the end of the third year, 4 sun (12 centimeters) at the end of the fourth year, and 5 sun at the end of the fifth year.

Wakin

In the case of the wakin, twice as many as of the preceding variety are cultivated in a pond. When put into a miniature pond they are most lively, but do not live long.

Artificial Coloration

Various designs are artificially produced on the back of goldfish at Koriyama. Dilute hydrochloric acid is applied to the part where certain figures are desired to be produced. But the scales along the margin of the intended figures (such as badges or flowery patterns) being but partially colored, the results are not very satisfactory. This artificial coloration is best attempted in August or September, in the early morning. When the fish are purely red, the discoloration makes them very unsightly; besides, the color of the head can not be changed. For the purpose of artificial coloring, the water of the pond in which the fish are kept should in the first place be completely renewed and then they should be abundantly supplied with food. When they grow plump and fleshy the figures are put on. This practice has been known from remote times.

Interest and Value of Goldfish to the Japanese

On account of its beautiful form, its fine bright color, and graceful attractive motion when swimming, the goldfish has been for hundreds of years a great favorite with the people of Japan, and now different varieties are cultivated almost all over the Empire. In Hokkaido it is very difficult for the fish to survive the winter, owing to the intensely cold climate, and they are hardly ever hatched and cultivated there. Yet the favor with which they are regarded is extremely great and they are yearly brought there from Tokyo without number in the beginning of summer.

Such is the interest with which goldfish are regarded in Japan, and they are, moreover, admired by every class of the Japanese people. Some are kept in an artificial lake in a garden, some in tanks of various forms, others suspended in a glass globe from a ceiling, still others put into the pond of a miniature garden, and so forth. The kind and quality of the fish naturally vary with the class of people by whom they are kept. Those kept by persons of wealth and position are superior breeds specially selected, while people of limited means, holding in regard healthy and strong ones, prefer the wakin and the ryukin to others. Those enjoyed by children are mostly what are called "dregs" and often sold on fête days, which are many in Tokyo and other towns.

Although the goldfish is so extensively cultivated in Japan, as has been stated above, yet its value from a commercial standpoint can not be said to be very great. The inferior breeds are sold at 1 sen (half a cent) a piece, while the superior ones fetch a price of only 50 yen ($25) a pair. On rare occasions, however, 200 or 300 yen are paid for a single pair. In costliness the ranchu ranks first, the superior breeds of the oranda shishigashira are also of great value, and the ryukin comes next. The number of the last variety yearly exported from Yokohama and Kobe is great.

As already mentioned, the ranchu is regarded with special favor in Tokyo. An exhibition of this variety is held there every year in autumn by lovers of the fish, for the purpose of having the merits of their exhibits determined, and a successful breeder to whom an award of merit is made prides himself upon it. The exhibition lasts two days, on the first of which are examined the grown-up breeds and on the second the young in the first year. Being developed in color and form, the former naturally attest the extent of skill in the breeders and their value can be known at a glance. The latter are those hatched only in the preceding spring and as yet little developed in every respect; but these, after all the care lavished upon them by breeders, are to appear again fully grown up for contest at a future show, and on that account are full of interest and promise. The examination of fish in an exhibition is made in a shallow tub containing a white earthenware plate in the center. Two ranchu being placed in the plate are examined by connoisseurs as to their shade, dapples, and the form of the tail and body. Those perfect in every respect are awarded the "first best," and a list of the exhibits made in the order of their merits is given to the public. Every time the classification is made amid a stormy debate by the examiners. No positive criteria exist to guide one in the examination of the ranchu. Nevertheless, those uniformly bright red are considered the best, so far as coloration is concerned. Those perfect in form, however uniformly white they may be, are counted tolerably good. The variegated ones are generally unpopular; but in Osaka and its vicinity, those having fine dapples are greatly appreciated, especially if the head is of a bright red color. A ranchu having either a white body and bright red fins and mouth, or a bright red in color in both cheeks, is also admired. Every one of the breeds exhibited has its own name, which is given in the aforesaid list with that of the owner. Those who participate in the show are mostly nobles, wealthy merchants, and others in comfortable circumstances. On such occasions the very best breed fetches a price of 200 or 300 yen, but not one in ten thousand commands such a high price. The exhibition takes place chiefly in Tokyo but it has recently come to be held in Osaka also. Not a few goldfish breeders with fish of their own culture now come from localities lying far beyond Hakone to take part in a Tokyo exhibition. The reason why the exhibition is held in autumn is that the goldfish puts on the most brilliant colors in that season.

Literature Cited

MATSUBARA, S. 1906. The culture of fish and other water products in Japan.—Proceedings of the Third International Fishery Congress. Vienna, 1905. Goldfish, pages 314–318, 1 colored plate with 5 figures.

MATSUBARA, S. 1909. Goldfish and their culture in Japan.—Proceedings of the Fourth International Fishery Congress, Washington, 1908.

MITSUKURI, K. 1905. The cultivation of marine and freshwater animals in Japan.—Bulletin of the U. S. Bureau of Fisheries for 1904, pages 257–289. The goldfish, pages 266–275, 4 plates.

RYDER, JOHN A. 1893. The inheritance of modifications due to disturbances of the early stages of development, especially in the Japanese domesticated races of gold-carp.—Proceedings of the Academy of Natural Sciences of Philadelphia, 1893, pages 75–94.

WATASE, S. 1887. On the caudal and anal fins of goldfish.— Journal of the Science College, Imperial University, Tokyo, Japan, vol. I, pages 247–267, 3 plates.

About the Author

Hugh M. Smith was United States Deputy Commissioner of Fisheries as well as President of the American Fisheries Society when this book was published in 1909. From 1907–8 he was Secretary-General of the Fourth International Fishery Congress; in 1908 Fellow of the American Association for the Advancement of Science. Member of the Washington Academy of Sciences, Biological Society of Washington, etc. Honorary Member of the Imperial and Royal Austrian Fishery Society, the Imperial Russian Society of Fish Culture and Fishing, the Salmon and Trout Association of Great Britain, and Corresponding Member of the German Sea Fishery Society.